# COSMOLOGY + 1

Readings from
**SCIENTIFIC
AMERICAN**

# COSMOLOGY + 1

With *Introduction by*
Owen Gingerich

*Harvard University*
*and*
*Smithsonian Astrophysical Observatory*

SAO
 W. H. Freeman and Company
San Francisco

COVER ILLUSTRATION: A synthetic optical photograph of our
Galaxy as it might appear to an observer in a distant external galaxy.
Produced by S. Christian Simonson III of the University of Maryland.
Permission for reproduction granted by A. Blaauw, Chairman, Board of
Directors, *Astronomy & Astrophysics*.

**Library of Congress Cataloging in Publication Data**
Main entry under title:

Cosmology + 1.

   Includes bibliographies and index.
   1. Cosmology—Addresses, essays, lectures.  2. Stars—
Evolution—Addresses, essays, lectures.  I. Gingerich,
Owen.  II. Scientific American.
QB981.C823    523.1'08                      77-1448
ISBN 0-7167-0012-3
ISBN 0-7167-0011-5 pbk.

Printed in the United States of America

9 8 7 6 5 4 3 2 1

# CONTENTS

*Note on cross-references to* SCIENTIFIC AMERICAN *articles:* Articles included in this book are referred to by title and page number; articles not included in this book but available as Offprints are referred to by title and offprint number; articles not included in this book and not available as Offprints are referred to by title and date of publication.

# PREFACE

Infinity, curved space, the big bang, red shifts of galaxies—these are the makings of modern cosmology. Armed with sophisticated mathematics and all-too-sparse observations, cosmologists have labored since the 1920s to construct a picture of the universe on its grandest scale. But the unfettered imaginations of yesterday's thinkers scarcely anticipated the surprises that modern observational science has brought forth: cataclysmic galactic nuclei, quasars, gigantic radio galaxies. Enormous in size and energy compared to our terrestrial station in space, yet only small units in the cosmos at large, these entities offer fresh clues to the violent history of the expanding universe. Perhaps someday they will help solve the mystery of how the explosion of space and energy began, and whether it will ever take place again.

This collection of articles from *Scientific American* describes various aspects of our search for an understanding of the universe as a whole. It includes descriptions of galaxies, of the pervasive background radiation, and of non-Euclidian space. It also includes two recent articles on black holes, those astonishing space-warps in miniature that may ultimately yield crucial hints about the curvature of space as a whole.

Is the universe a one-time happening? Or does it pulse in an unending cycle? This is today the greatest question of cosmology, but to call it the greatest question of all is to confuse size with significance. There are indeed many other important inquiries that challenge the imagination, and within the wide scope of astronomy there is one rival in particular: Are we alone? Are we the only intelligent life in this vast celestial frame? For this reason this collection is called *Cosmology + 1:* the extra article concerns the fascinating but so far unsuccessful search for life elsewhere. Like the question whether our universe has exploded in a one-and-only incarnation, the question whether intelligent beings inhabit the satellites of some distant stars is quite undecided and has called forth wild speculation. These questions are intellectual mainsprings that drive much astronomical research today. I hope the reader will find the search for their answers exciting, even though they may never be found.

*January 1977*                                                    *Owen Gingerich*

COSMOLOGY + 1

# COSMOLOGY + 1

## INTRODUCTION

With the publication of Newton's *Principia* in 1687, the stage was set for a new kind of inquiry into the nature of the universe on its largest scale. Given the concept of Universal Gravitation, the obvious question was Why didn't all the stars draw themselves together into one grand fiery ball? The solution to this puzzle proved so elusive that cosmology simply went into hibernation for over two centuries.

Today, the idea of gravitational attraction leading to gravitational collapse plays a key role in our understanding of many astronomical phenomena. The sun was once an extended gaseous sphere as large as the present solar system. Warming as it shrank, it finally achieved sufficiently high internal temperatures to trigger nuclear reactions, which have temporarily balanced the powerful gravitational pull. Eventually, when these fuels have been exhausted, gravity will crush the sun's material into a smaller, denser ball.

Elsewhere, gravitational forces have drawn gaseous spheres that were more massive (and hence have evolved more rapidly) into dense neutron stars. Some of these betray their existence by projecting radio beams that sweep across the earth as the neutron stars spin. Called pulsars, these objects were first detected only about a decade ago. Objects still more gravitationally compressed, the black holes, have become the subject of intense speculation. Black holes may have been detected in a few unusual binary systems and in the centers of some globular clusters.

Our Milky Way galaxy, too, shows signs of gravitational collapse. Initially its mass was spread throughout a giant sphere, and today the distribution of globular clusters and isolated stars is a visible vestige of our galaxy's earlier shape. Most of the original gas, dust, and stars has now been pulled into a rotating pinwheel about 100,000 light-years across. As in the formation of the solar system, rotational momentum prevented these objects from falling directly into the center of the galaxy, which is why the stars remain distributed in a flat plane rather than in a small central conglomeration.

But what about the universe as a whole? Will it not also collapse under the inexorable gravitational tug? Newton's question has been revived to become the leading problem of cosmology today. In modern terms, astronomers ask if the universe is open or closed. If space is hyperbolic, then the universe is open and unbounded, and the galaxies will forever rush away from one another, leading to an ever colder, fainter, and more tenuous distribution of matter. On the other hand, if space is spherical, then the universe is closed and bounded, and its expansion will eventually slow to a stop, followed by contraction and a mighty implosion. The open universe is a one-time affair, but the closed universe might be a single cycle of an infinite series of oscillations.

For several decades in this century the only apparent way to answer the question whether the universe is open or closed was to examine the red shifts of the distant galaxies. Hence A. R. Sandage's 1956 article, "The Red-Shift," leads off this collection, followed by G. Gamow's ("The Evolutionary Universe") and J. J. Callahan's ("The Curvature of Space in a Finite Universe") articles on the curvature of space. In all of the simple evolutionary models, the universe is gradually slowing down in its expansion; for closed space the deceleration will bring the expansion to a stop, but for open space the smaller retardation is insufficient to stop the outward rush of the galaxies. In both cases, the remote galaxies, whose light started out in the distant past, should show signs of the greater speeds at the earlier epoch. If the universe is slowing down enough to come to a stop, then these galaxies should be rushing away from us even faster than if the universe is destined to expand indefinitely. Subtle measurements are required. For very faint galaxies, which are seen at a much younger stage because they are so far away, evolutionary effects must be taken into account. Are younger galaxies intrinsically more luminous? Probably, unless separate galaxies sometimes coalesce into one as they grow older. As yet the theory is too uncertain to allow the relative distances to be established with enough precision to settle the openness or closure of the universe by deceleration observations.

If the absolute distance of the great cloud of galaxies in Virgo could be determined, then their red shifts, in combination with accurate age determinations from the relative abundances of nuclear isotopes, would help decide the question. The red-shift/distance relation ("Hubble's constant"), when directly extrapolated back to the point of zero expansion, yields an age for the universe ranging from 8 billion to 20 billion years, depending on the distance chosen for the Virgo galaxies. Because the galaxies rushed apart more rapidly in the earlier stages of the expansion, the actual age of the universe would be less than $\frac{2}{3}$ of the directly extrapolated age if the universe were closed, and somewhat more than $\frac{2}{3}$ if the universe were open. Thus, for example, if the direct extrapolation gave an age for the universe of 15 billion years, and an examination of the abundances of radioactive rhenium isotopes gave an age for the universe of 12 billion years, then the universe would be open, because 12 billion is greater than $\frac{2}{3}$ of 15 billion. The difficulty lies in finding the distances to the galaxies, for the measurements have proved to be shifting and unreliable. For example, Sandage's illustration on p. 6 gives the distance of the galaxy in Virgo as 22 million light-years; in the past decade, a distance twice as great has been commonly accepted; and Sandage's most recent determination is about 70 million light-years. The other distances in the figure must be adjusted correspondingly. At present, this procedure is as unreliable as establishing the deceleration from observations of the more distant galaxies.

Another approach to the problem is to determine the mean density of the universe. If this exceeds the critical value of about $5 \times 10^{-30}$ gm/cm$^3$, then there is sufficient gravitating mass to pull the universe back together again. Otherwise it will be Humpty Dumpty on a grand scale. (The value of the critical density depends on the expansion rate, but so does the determination of the mean density; consequently, a recalibration of intergalactic distances and the expansion rate would not affect this argument.) But ever since the 1920s, when Edwin Hubble first calculated the mean density of the universe, no one has ever found enough material to close space. Proponents of a closed universe have therefore been obliged to speculate on the whereabouts of the "missing mass."

The procedure for calculating the mean density of the universe depends on knowing the masses of various kinds of galaxies; the masses, in turn, are generally determined from the luminosities of these galaxies. However, if there is a considerable mass not accounted for by luminous objects—for example, multitudes of planets, black holes, or large gaseous extensions of the spiral disks—

then the total masses could be considerably underestimated. That this must be the case is strongly suggested by the individual motions of galaxies in clusters. These motions are sufficiently large that the galaxies in the clusters would have drifted apart unless the clusters contain invisible additional gravitating matter. Unfortunately for the advocates of a closed universe, raising the galaxy masses enough to stabilize the clusters is still not quite sufficient to raise the density of the universe to the critical value.

It is possible that the "missing mass" is in the form of hot gas between the galaxy clusters themselves. Recent calculations show that atoms heated to nearly a billion degrees Kelvin by exploding galaxies would produce X-radiation consistent with current observations. Perhaps, as Aristotle taught, nature really does abhor a vacuum, and enough highly ionized atoms will be found in intergalactic space to close the universe!

Should sufficient matter be found to raise the calculated mean density to $5 \times 10^{-30}$ gm/cm$^3$, however, a serious conflict then would arise with predictions based on the cosmic abundance of deuterium. Astronomers presently believe that the deuterium was formed in the earliest stages of the big bang: the lower the present density of the universe, the more deuterium must have been formed at the beginning. J. R. Gott, J. E. Gunn, D. N. Schramm, and B. M. Tinsley deal cogently with this rather intricate reasoning in their article, "Will the Universe Expand Forever?" but as they admit, the details of the theory are still somewhat uncertain. Thus, although a number of arguments support the concept of an open universe, the outcome is far from settled.

Between the selections on the curvature of space and the summary arguments in favor of an open universe by Gott, Gunn, Schramm, and Tinsley there are two groups of articles. The first group concerns characteristics of the universe as a whole. D. Sciama, in "Cosmology before and after Quasars," a book review, discusses the demise of the steady-state cosmology. Then follow detailed articles on the two discoveries that brought the big-bang cosmology into ascendancy: A. Webster, "The Cosmic Background Radiation," and M. Schmidt and F. Bello, "The Evolution of Quasars." Finally, M. J. Rees and J. Silk, in "The Origin of Galaxies," relate the formation of galaxies to the evolution of the universe as a whole.

The second group deals not with the universe in the large, but with a small component whose nature may provide important insights into the structure of the universe itself. K. S. Thorne writes convincingly on "The Search for Black Holes," and S. W. Hawking, in "The Quantum Mechanics of Black Holes," describes some of the unexpected properties of bulk matter in its most compact configuration.

Like the articles on cosmology, the "+ 1" selection, "The Search for Extraterrestrial Intelligence," by C. Sagan and F. Drake, grapples with a speculative and conceivably large-scale characteristic of the universe. The possibility of detecting an extraterrestrial civilization has been wrested from science fiction and has become serious science. This is not to say that the quest finds enthusiastic support in all quarters. The chain of reasoning leading to a high probability that intelligent life exists elsewhere is no doubt even more tenuous than the cosmological dream castles of half a century ago. Nevertheless, the search is exciting and the arguments for its pursuit deserve a wider discussion. Here is a place to begin!

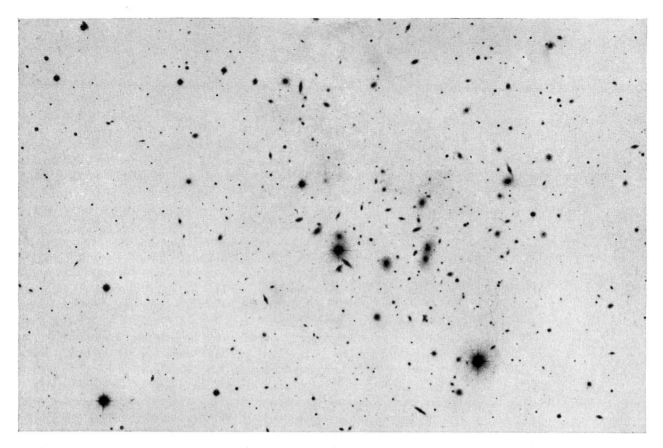

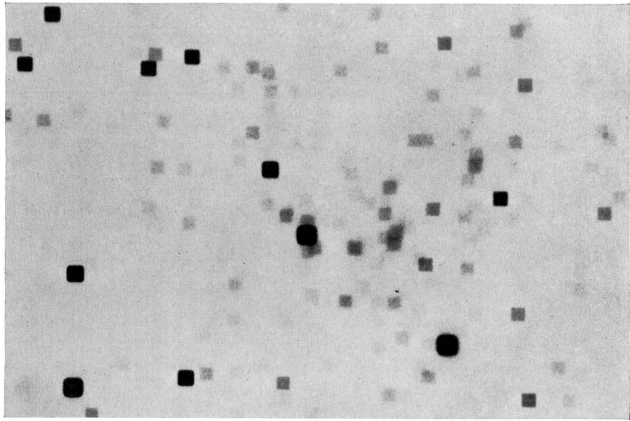

BRIGHTNESS OF GALAXIES may be measured with the help of the jiggle camera (*see photograph on the opposite page*). At the top is a negative print of a 200-inch telescope photograph showing nearby stars and a cluster of galaxies in Corona Borealis. Although the stars have made the brightest images in the photograph, they are essentially point sources of light. The galaxies, on the other hand, are extended sources of light. To measure the brightness of a galaxy by comparing it with the known brightness of a star, the two images must be made the same. This is done by smearing the images as shown at bottom in a jiggle-camera photograph of the same area.

# The Red-Shift

by Allan R. Sandage
September 1956

*The redness, and presumably the speed of recession, of most galaxies increases regularly with distance. The most distant galaxies observed appear to depart from this law, a fact of deep meaning for cosmology*

In the nature of things it is a delicate undertaking to try to discern the general structure and features of a universe which stretches out farther than we can see. For more than a quarter of a century both the theoreticians and the observers of the cosmos have been making exciting discoveries, but the points of contact between the discoveries have been few. The predictions of the theorists, deduced from the most general laws of physics, are not easy to test against the real world—or rather, the small portion of the real world that we can observe. There is, however, one solid meeting ground between the theories and the observations, and that is the apparent expansion of the universe. Other aspects of the universe may be interpreted in different ways to fit different theories, but concerning the expansion the rival theories make unambiguous predictions on which they will stand or fall. There is now hope that red-shift measurements of the universe's expansion with the 200-inch telescope on Palomar Mountain will soon make it possible to decide, among other things, whether we live in an evolving or a steady-state universe.

Let us begin by considering just what issue the measurements seek to decide, as set forth by Gamow ("The Evolutionary Universe," p. 12) and Hoyle ("The Steady-State Universe, September 1956). The steady-state theory says that the universe has been expanding at a constant rate throughout an infinity of time. The evolutionary theory, in contrast, implies that the expansion of the universe is steadily slowing down. If the universe began with an explosion from a superdense state, its rate of expansion was greatest at the beginning and has been slowing ever since because of the opposing gravitational attraction of its matter, which acts as a brake on the expansion—much as an anchored elastic string attached to a golf ball would act as a brake on the flight of the ball.

Now in principle we can decide whether the rate of expansion has changed or not simply by measuring the speed of expansion at different times in the universe's history. And the 200-inch telescope permits us to do this. It covers a range of about two billion years in time. We see the nearest galaxies as they were only a few million years ago, while the light from the most distant galaxies takes so long to reach us that we see them at a stage in the universe's history going back to one or two billion years ago. If the explosion theory is correct, the universe should have been expanding at an appreciably faster rate then than it is now. Since the light we are receiving from the distant galaxies is a flashback to that earlier time, its red-shift should show them receding from us faster than if the rate of expansion had remained constant.

The red-shift is so basic a tool for testing our notions about the universe that it is worthwhile to review how it was discovered and how it is used.

An astronomer cannot perform experi-

**JIGGLE CAMERA** smears the images by moving the photographic plate in a rectangle during exposure. It is mounted in the prime-focus cage at the upper end of the 200-inch.

ments on the objects of his study, or even examine them at first hand. All his information rides on beams of light from outer space. By sufficiently ingenious instruments and equally ingenious interpretation (we hope), he may translate this light into information about the temperatures, sizes, structures and motions of the celestial bodies. It was in 1888 that a German astronomer, H. C. Vogel, first demonstrated that the spectra of stars could give information about motions which could not otherwise be detected. He discovered the Doppler effect in starlight.

The Doppler effect, as every physics student knows, is a change in wavelength observable when the source of radiation (sound, light, etc.) is in motion. If it is moving toward the observer, the wavelength is shortened; if away, the waves are lengthened. In the case of a star moving away from us, the whole spectrum of its light is shifted toward the red, or long-wave, end.

This spectrum, made by means of a prism or diffraction grating which spreads the light out into a band of its component colors, is usually not continuous. Certain wavelengths of the light are absorbed by atoms in the star's atmosphere. For example, most stars show strong absorption, by calcium atoms, at the wavelengths of 3933.664 and 3968.-470 Angstrom units. (An Angstrom unit is a hundred-millionth of a centimeter.) The absorption is signaled by dark lines in the spectrum, known in this case as the K and H lines of calcium. Now if a star is moving away from us, these lines will be displaced toward the red end of the spectrum. In the spectrum of the star known as Delta Leporis, for in-

RED-SHIFT of four galaxies on this page is depicted in the spectra on the opposite page. The galaxies are centered in the photographs. The spectra are the bright horizontal streaks tapered to the left and right. Above and below each spectrum are comparison lines from the spectrum of iron. Near the left end of the spectrum at the top of the page are two dark vertical lines: the K and H lines of calcium. If the galaxy did not exhibit the red-shift, these lines would be in the position of the broken line running vertically down the page. The amount of their shift toward the red, or right, end of the spectrum is indicated by the short arrow to the right of the broken line. The larger shift of the K and H lines of the three fainter galaxies is indicated by the longer arrows below their spectra. The constellation, approximate distance and velocity of recession of each galaxy is at left of its photograph.

VIRGO

22 MILLION LIGHT-YEARS

1,200 KILOMETERS PER SECOND

CORONA BOREALIS

400 MILLION LIGHT-YEARS

21,500 KILOMETERS PER SECOND

BOOTES

700 MILLION LIGHT-YEARS

39,300 KILOMETERS PER SECOND

HYDRA

1.1 BILLION LIGHT-YEARS

60,900 KILOMETERS PER SECOND

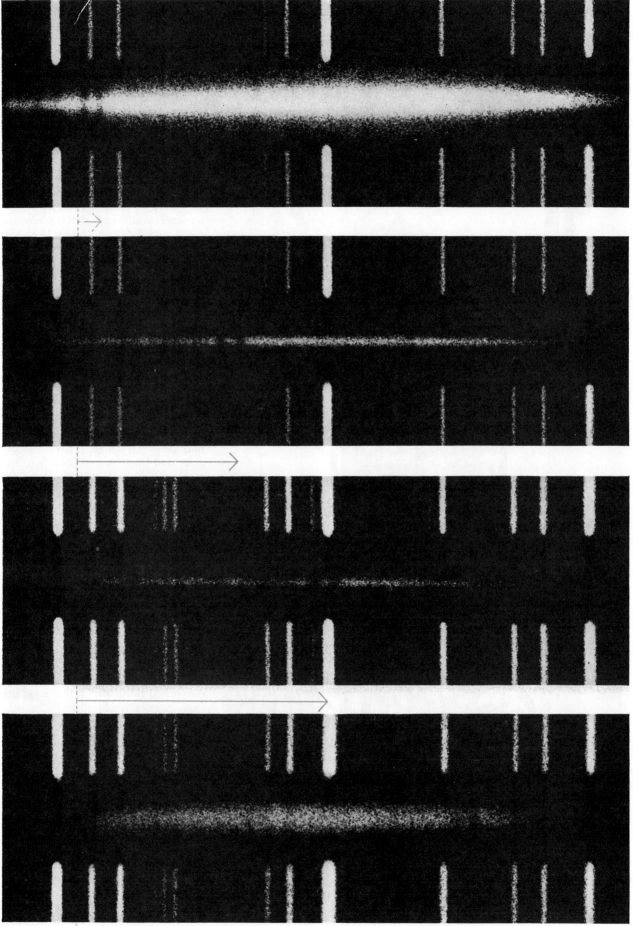

stance, the K line of calcium is displaced 1.298 Angstroms toward the red. Assuming the displacement is due to the Doppler effect, it is a simple matter to calculate the velocity of the star's receding motion. Dividing the amount of the displacement by the normal wavelength at rest, and multiplying by the speed of light (300,000 kilometers per second) we get the speed of the star—in this case 99 kilometers per second. The calculation on the basis of displacement of the H line gives the same figure.

Equipped with this powerful tool,

SPECTRA ARE MADE with the spectrograph at the top, which is mounted at the prime focus of the 200-inch telescope. Inside the spectrograph the converging rays of the 200-inch mirror are made parallel by a concave mirror. The light is then dispersed by a diffraction grating. At the bottom is a Schmidt camera used to photograph the spectrum. It has an optical path of solid glass and a speed of $f/.48$. The plateholder and plate are below the camera.

many of the large observatories in the world spent a major part of their time during the early part of this century measuring the velocities of receding and approaching stars in our galaxy. At first it was a work of pure curiosity, no one suspecting that it might have any bearing on cosmological theories. But in the 1920s V. M. Slipher of the Lowell Observatory made a discovery which was to lead to a completely new picture of the universe. His measurements of redshifts of a number of "nebulae" then thought to lie in our galaxy showed that they were all receding from us at phenomenal speeds—up to 1,800 kilometers per second. Edwin P. Hubble at Mount Wilson soon established that the "nebulae" were systems of stars, and he went on to measure their distances. The method he used was the one developed by Harlow Shapley, employing Cepheid variable stars as the yardstick. Shapley had found a way to measure the intrinsic brightness of these stars, and therefore their distance could be estimated from their apparent brightness by means of the rule that the intensity of light falls off as the square of the distance. Hubble observed that the galaxies nearest our own system, including the Great Nebula in Andromeda, contained Cepheid variables, and when he computed their distances he came out with the then astounding figure of about one million light-years! He next tackled the problem of finding the distances of Slipher's nebulae. Since variable stars could not be detected in them, he used their brightest stars as distance indicators instead. He found that these nebulae were at distances ranging up to 20 million light-years from us, and what was more remarkable, their velocities increased in strict proportion to their distances!

Hubble made the daring conjecture that the universe as a whole was expanding. He predicted that more remote galaxies would show larger redshifts, still in proportion to their distance. To test Hubble's speculation, Milton L. Humason began a long-range program of spectral analysis of more distant galaxies with the 100-inch telescope on Mount Wilson. In these faint galaxies it was no longer possible to distinguish even the bright stars, and so the relative brightness of the galaxy as a whole had to be taken as the measure of distance. That is, a galaxy one fourth as bright as another was assumed to be twice as far away. Hubble reasoned that while individual galaxies might deviate from this rule, statistically the population of galaxies as a whole would follow it. The prin-

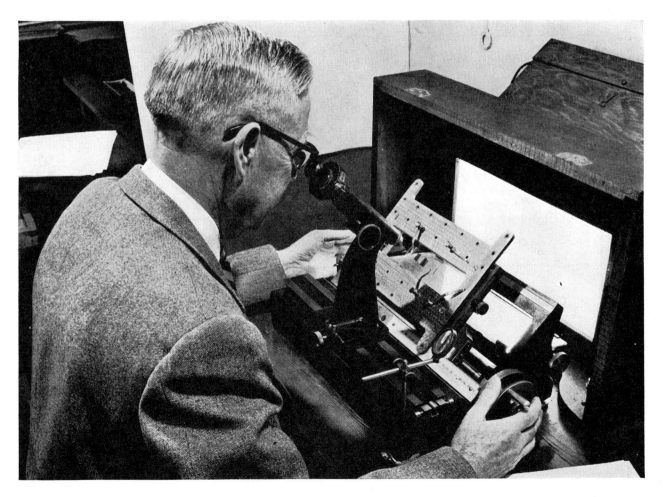

ACTUAL SIZE of the red-shift spectrum is indicated by the photograph at the top of the page. The glass photographic plate is 15 millimeters on an edge. The spectrum is 5 mm. long. At bottom Milton L. Humason examines a spectrum with a low-power microscope.

ciple is still the basis of distance determinations today.

Humason laboriously photographed spectra of galaxies, and Hubble measured their apparent brightness, from 1928 to 1936, when they reached the limit of the 100-inch telescope. The history of the red-shift program in those years is a story of extreme skill and patience at the telescope and of steady improvement in instrumentation. It was a long and difficult task to photograph spectra then; the prisms used required long exposures, and it took 10 nights or more to obtain a spectrum which with modern equipment can be recorded in less than an hour today. The improvement in equipment includes not only the 200-inch telescope but also diffraction

gratings, faster cameras and a vast improvement in the sensitivity of photographic plates, thanks to the Eastman Kodak Company. Astronomers the world over, and cosmology, owe a large debt to the Eastman research laboratories.

Humason's first really big red-shift came early in 1928, when he got a spectrum of a galaxy called NGC 7619. Hubble had predicted that its velocity should be slightly less than 4,000 kilometers per second: Humason found it to be 3,800. By 1936, at the limit of the 100-inch telescope's reach, they had arrived at a cluster of galaxies, called Ursa Major No. 2, which showed a velocity of 40,-000 kilometers per second. All the way out to that range of more than half a billion light-years the velocity of galax-

ies increased in direct proportion to the distance. In a sense this was disappointing, because the various cosmological theories predicted that some change in this relation should begin to appear when the observations had been pushed far enough. Further exploration into the distances of space had to await the completion of the 200-inch Hale telescope on Palomar Mountain.

In 1951 the red-shift program was resumed, with a new spectrograph of great speed and versatility placed in the big telescope's prime focus cage, where the observer rides with his instruments. The spectrograph has to be of very compact design to fit into the cramped space of the cage. The photographic plate itself, mounted in the middle of a complex

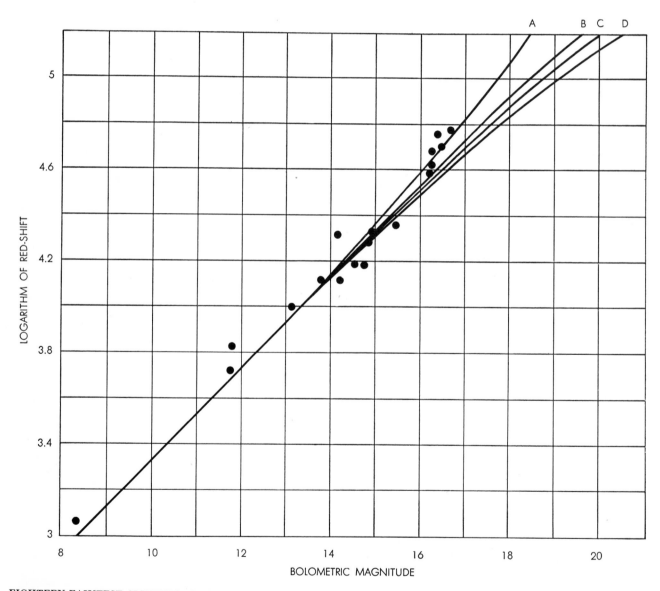

**EIGHTEEN FAINTEST CLUSTERS** of galaxies yet measured are plotted for their red-shift (or speed of recession) and apparent magnitude (or distance). Line C is a universe expanding forever at the same rate. Line D is a steady-state universe. If the line falls to the left of C, the expansion must slow down. If it falls between C and B, the universe is open and infinite. If it falls to the left of B, the universe is closed and finite. If it falls on B, it is Euclidean and infinite. A is the trend suggested by the six faintest clusters.

optical arrangement, is only 15 millimeters (about half an inch) on a side. The cutting and handling of such small pieces of glass in complete darkness (to avoid exposure of the plate) is a tricky business. The spectrum recorded on the plate is a tiny strip only a fifth of an inch long, but it is long enough to measure red-shifts to an accuracy of better than one half of 1 per cent.

The most distant photographable galaxies are so faint that they are not visible to the eye through the telescope: they can be recorded only by extended exposure of the plate. The observer guiding the telescope must position the slit of the spectrograph by reference to guide stars within the same field as the distant object. Another great difficulty in recording the red-shift of extremely distant galaxies arises from the magnitude of the shift. The displacement of the calcium dark lines toward the red is so large that the lines move clean off the sensitive range of blue photographic plates, which astronomers like to use because of their speed. So slow panchromatic plates must be used, and Humason has been forced to return to exposure times as long as 30 hours or more.

The other part of the program—measuring the distances of the galaxies—also has been helped by improvements in technique. For measurement of their brightness the Mount Wilson telescopes employ photomultiplier tubes, which amplify the light energy by electronic means. Such equipment was not available for the 200-inch telescope when the present program began. Instead the intensity of the light from very faint galaxies was measured by a tricky method which compares it with that of stars of known magnitude. No direct comparison can be made, of course, between the picture of a star and that of a galaxy or cluster of galaxies, because the star is a point source of light while a galactic system is a spread-out image. To make the images comparable, a region of the sky is photographed with a "jiggle" camera which moves the plate around so that the images of stars and of galaxies are smeared out in squares [see photograph at bottom of page 4]. They can then be compared as to brightness—just as one may use color cards to find a match to the color of a room.

Humason has now measured red-shifts of remote clusters of galaxies with recession velocities up to 60,000 kilometers per second. What do they show? Is the velocity still increasing in strict proportion to the distance?–

The information about 18 of the faintest measured clusters is given in the accompanying chart [see page 10]. Their velocities are plotted against their apparent brightness, or estimated distances. If velocity increases in direct proportion to the distance, the observed velocity-distance relation should be "linear" (i.e., follow a straight line). But as the chart shows, the very faintest clusters have begun to depart from that line. These clusters, about a billion light-years away, are moving faster (by about 10,000 kilometers per second) than in direct proportion to their apparent distance. In other words, the data would be interpreted to mean that a billion years ago the universe was expanding faster than it is now. If the measurements and the interpretation are correct, this suggests that we live in an evolving rather than in a steady-state universe.

The observed change in the curve buys us much more information. To begin with, it tells us something about the mean density of matter in the universe. The rate at which the expansion of the universe is slowing down (if it is) depends on the mean density of its matter: the higher the density, the greater the braking effect. The amount of departure from linearity indicated by the measurements thus far calls for a mean density of about 3 x 10$^{-28}$ grams of matter per cubic centimeter (about one hydrogen atom per five quarts of space). Now this amounts to about 300 times the total mass of the matter estimated to be contained in galaxies: that figure comes out to a mean density of only 10$^{-30}$ grams per cubic centimeter. If our present tentative value for the slowdown of the expansion should be confirmed, we would have to conclude that either the current estimates of the masses of the galaxies are wrong or that there is a great deal of matter, so far undetected, in intergalactic space. Matter in the form of neutral hydrogen (i.e., normal hydrogen atoms consisting of a proton and an electron) might be present in space and still have escaped detection until now because it is not luminous. The giant radio telescopes now under construction or on the drawing boards perhaps will detect the hydrogen, if it exists in the postulated quantities.

Once we know the rate at which expansion of the universe is slowing down, it becomes possible to determine not only the mean density of matter but also the geometry of space—that is, its curvature. Models of the evolving universe take three forms: the Euclidean case, in which

space is flat, open and infinite; a curved universe which is closed and finite, like the surface of a sphere; and a curved universe which is open and infinite, like the surface of a saddle. In the accompanying velocity-distance chart [page 10] curves to the left of C represent evolving models, and curve D represents the steady-state model. If the curve of the velocity-distance relation lies between C and B, the universe is open and infinite. Line B is the Euclidean case of flat space. If the curve is left of B, the universe is closed and finite, the radius of its curvature decreasing as we move farther to the left.

According to our present observations, the actual relation follows a curve left of B (curve A on the chart). Although our data are still crude and inconclusive, they do suggest that the steady-state model does not fit the real world, and that we live in a closed, evolving universe.

Humason has gone beyond 60,000 kilometers per second and attempted to measure the red-shifts of two faint clusters whose predicted velocity is more than 100,000 kilometers per second. So far these efforts have not yielded reliable results, but he is continuing them. These two remote clusters may well hold the key to the structure of the universe. We stand a chance of finding the answer to the cosmological problem. The red-shift program will continue toward this goal.

If the expansion of the universe is decelerating at the rate our present data suggest, the expansion will eventually stop and contraction will begin. If it returns to a superdense state and explodes again, then in the next cycle of oscillation, some 15 billion years hence, we may all find ourselves again pursuing our present tasks.

Although no final answers have yet emerged, big steps have been taken since 1928 toward the solution to the cosmological problem, and there is hope that it may now be within our grasp. The situation has nowhere been better expressed than in Hubble's last paper:

"For I can end as I began. From our home on the earth we look out into the distances and strive to imagine the sort of world into which we are born. Today we have reached far out into space. Our immediate neighborhood we know rather intimately. But with increasing distance our knowledge fades . . . until at the last dim horizon we search among ghostly errors of observations for landmarks that are scarcely more substantial. The search will continue. The urge is older than history. It is not satisfied and it will not be suppressed."

# 2

# The Evolutionary Universe

by George Gamow
*September 1956*

*Most cosmologists believe that the universe began as a
dense kernel of matter and radiant energy which
started to expand about five billion years ago and
later coalesced into galaxies*

Cosmology is the study of the general nature of the universe in space and in time—what it is now, what it was in the past and what it is likely to be in the future. Since the only forces at work between the galaxies that make up the material universe are the forces of gravity, the cosmological problem is closely connected with the theory of gravitation, in particular with its modern version as comprised in Albert Einstein's general theory of relativity. In the frame of this theory the properties of space, time and gravitation are merged into one harmonious and elegant picture.

The basic cosmological notion of general relativity grew out of the work of great mathematicians of the 19th century. In the middle of the last century two inquisitive mathematical minds—a Russian named Nikolai Lobachevski and a Hungarian named János Bolyai—discovered that the classical geometry of Euclid was not the only possible geometry: in fact, they succeeded in constructing a geometry which was fully as logical and self-consistent as the Euclidean. They began by overthrowing Euclid's axiom about parallel lines: namely, that only one parallel to a given straight line can be drawn through a point not on that line. Lobachevski and Bolyai both conceived a system of geometry in which a great number of lines parallel to a given line could be drawn through a point outside the line.

To illustrate the differences between Euclidean geometry and their non-Euclidean system it is simplest to consider just two dimensions—that is, the geometry of surfaces. In our schoolbooks this is known as "plane geometry," because the Euclidean surface is a flat surface. Suppose, now, we examine the properties of a two-dimensional geometry constructed not on a plane surface but on a curved surface. For the system of Lobachevski and Bolyai we must take the curvature of the surface to be "negative," which means that the curvature is not like that of the surface of a sphere but like that of a saddle [*see illustrations on page 14*]. Now if we are to draw parallel lines or any figure (*e.g.,* a triangle) on this surface, we must decide first of all how we shall define a "straight line," equivalent to the straight line of plane geometry. The most reasonable definition of a straight line in Euclidean geometry is that it is the path of the shortest distance between two points. On a curved surface the line, so defined, becomes a curved line known as a "geodesic" [see "The Straight Line," by Morris Kline; SCIENTIFIC AMERICAN, March 1956].

Considering a surface curved like a saddle, we find that, given a "straight" line or geodesic, we can draw through a point outside that line a great many geodesics which will never intersect the given line, no matter how far they are extended. They are therefore parallel to it, by the definition of parallel. The possible parallels to the line fall within certain limits, indicated by the intersecting

*Five contributors to modern cosmology are depicted in these drawings by Bernarda Bryson.*

lines in the drawing at the left in the middle of the next page.

As a consequence of the overthrow of Euclid's axiom on parallel lines, many of his theorems are demolished in the new geometry. For example, the Euclidean theorem that the sum of the three angles of a triangle is 180 degrees no longer holds on a curved surface. On the saddle-shaped surface the angles of a triangle formed by three geodesics always add up to less than 180 degrees, the actual sum depending on the size of the triangle. Further, a circle on the saddle surface does not have the same properties as a circle in plane geometry. On a flat surface the circumference of a circle increases in proportion to the increase in diameter, and the area of a circle increases in proportion to the square of the increase in diameter. But on a saddle surface both the circumference and the area of a circle increase at *faster* rates than on a flat surface with increasing diameter.

After Lobachevski and Bolyai, the German mathematician Bernhard Riemann constructed another non-Euclidean geometry whose two-dimensional model is a surface of positive, rather than negative, curvature—that is, the surface of a sphere. In this case a geodesic line is simply a great circle around the sphere or a segment of such a circle, and since any two great circles must intersect at two points (the poles), there are no parallel lines at all in this geometry. Again the sum of the three angles of a triangle is not 180 degrees: in this case it is always *more* than 180. The circumference of a circle now increases at a rate *slower* than in proportion to its increase in diameter, and its area increases more slowly than the square of the diameter.

Now all this is not merely an exercise in abstract reasoning but bears directly on the geometry of the universe in which we live. Is the space of our universe "flat," as Euclid assumed, or is it curved negatively (per Lobachevski and Bolyai) or curved positively (Riemann)? If we were two-dimensional creatures living in a two-dimensional universe, we could tell whether we were living on a flat or a curved surface by studying the properties of triangles and circles drawn on that surface. Similarly as three-dimensional beings living in three-dimensional space we should be able, by studying geometrical properties of that space, to decide what the curvature of our space is. Riemann in fact developed mathematical formulas describing the properties of various kinds of curved space in three and more dimensions. In the early years of this century Einstein conceived the idea of the universe as a curved system in four dimensions, embodying time as the fourth dimension, and he proceeded to apply Riemann's formulas to test his idea.

Einstein showed that time can be considered a fourth coordinate supplementing the three coordinates of space. He connected space and time, thus establishing a "space-time continuum," by means of the speed of light as a link between time and space dimensions. However, recognizing that space and time are physically different entities, he employed the imaginary number $\sqrt{-1}$, or $i$, to express the unit of time mathematically and make the time coordinate formally equivalent to the three coordinates of space.

In his special theory of relativity Einstein made the geometry of the timespace continuum strictly Euclidean, that is, flat. The great idea that he introduced later in his general theory was that gravitation, whose effects had been neglected in the special theory, must make it curved. He saw that the gravitational effect of the masses distributed in space and moving in time was equivalent to curvature of the four-dimensional spacetime continuum. In place of the classical Newtonian statement that "the sun produces a field of forces which impels the earth to deviate from straight-line mo-

*From left to right they are: Nikolai Lobachevski, Bernhard Riemann, Albert Einstein, Willem de Sitter and Georges Lemaitre*

14

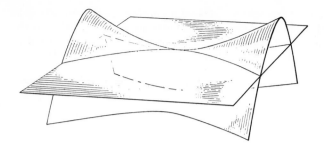

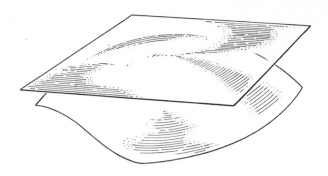

NEGATIVE AND POSITIVE CURVATURE of space is suggested by this two-dimensional analogy. The saddle-shaped surface at left, which lies on both sides of a tangential plane, is negatively curved.

The spherical surface at right, which lies on one side of a tangential plane, is positively curved. If space is negatively curved, the universe is infinite; if it is positively curved, the universe is finite.

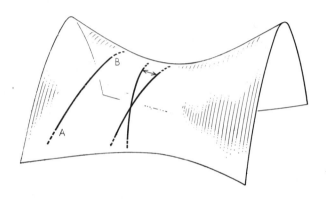

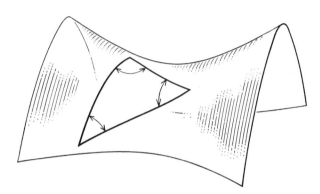

ON A NEGATIVELY CURVED SURFACE the shortest distance between two points is not a straight line but a curved "geodesic," such as the line AB at the left. On a plane surface only one parallel to a given straight line can be drawn through a point not on that line; on a negatively curved surface many geodesics can be drawn

through a point not on a given geodesic without ever intersecting it. These "parallel" lines will fall within the limits indicated by the arrow between the intersecting lines at left. On a plane surface the angles of a triangle add up to 180 degrees; on the negatively curved surface at the right, they add up to less than 180 degrees.

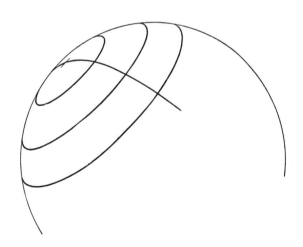

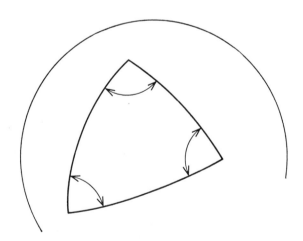

ON A POSITIVELY CURVED SURFACE the shortest distance between two points follows a great circle, a closed line passing through opposite points on the surface (single curved line at left). In this geometry there are no "parallel" lines because any two

great circles must intersect. The circumference of a circle increases more slowly with diameter than on a flat surface, and the area similarly increases more slowly (concentric circles at left). The angles of a triangle on the surface (right) add up to more than 180 degrees.

tion and to move in a circle around the sun," Einstein substituted a statement to the effect that "the presence of the sun causes a curvature of the space-time continuum in its neighborhood."

The motion of an object in the space-time continuum can be represented by a curve called the object's "world line." For example, the world line of the earth's travel around the sun in time is pictured in the drawing on this page. (Space must be represented here in only two dimensions; it would be impossible for a three-dimensional artist to draw the fourth dimension in this scheme, but since the orbit of the earth around the sun lies in a single plane, the omission is unimportant.) Einstein declared, in effect: "The world line of the earth is a geodesic in the curved four-dimensional space around the sun." In other words, the line ABCD in the drawing corresponds to the shortest *four-dimensional* distance between the position of the earth in January (at A) and its position in October (at D).

Einstein's idea of the gravitational curvature of space-time was, of course, triumphantly affirmed by the discovery of perturbations in the motion of Mercury at its closest approach to the sun and of the deflection of light rays by the sun's gravitational field. Einstein next attempted to apply the idea to the universe as a whole. Does it have a general curvature, similar to the local curvature in the sun's gravitational field? He now had to consider not a single center of gravitational force but countless centers of attraction in a universe full of matter concentrated in galaxies whose distribution fluctuates considerably from region to region in space. However, in the large-scale view the galaxies are spread fairly uniformly throughout space as far out as our biggest telescopes can see, and we can justifiably "smooth out" its matter to a general average (which comes to about one hydrogen atom per cubic meter). On this assumption the universe as a whole has a smooth general curvature.

But if the space of the universe is curved, what is the sign of this curvature? Is it positive, as in our two-dimensional analogy of the surface of a sphere, or is it negative, as in the case of a saddle surface? And, since we cannot consider space alone, how is this space curvature related to time?

Analyzing the pertinent mathematical equations, Einstein came to the conclusion that the curvature of space must be independent of time, *i.e.*, that the universe as a whole must be unchanging

(though it changes internally). However, he found to his surprise that there was no solution of the equations that would permit a static cosmos. To repair the situation, Einstein was forced to introduce an additional hypothesis which amounted to the assumption that a new kind of force was acting among the galaxies. This hypothetical force had to be independent of mass (being the same for an apple, the moon and the sun!) and to gain in strength with increasing distance between the interacting objects (as no other forces ever do in physics!).

Einstein's new force, called "cosmic repulsion," allowed two mathematical models of a static universe. One solution, which was worked out by Einstein him-

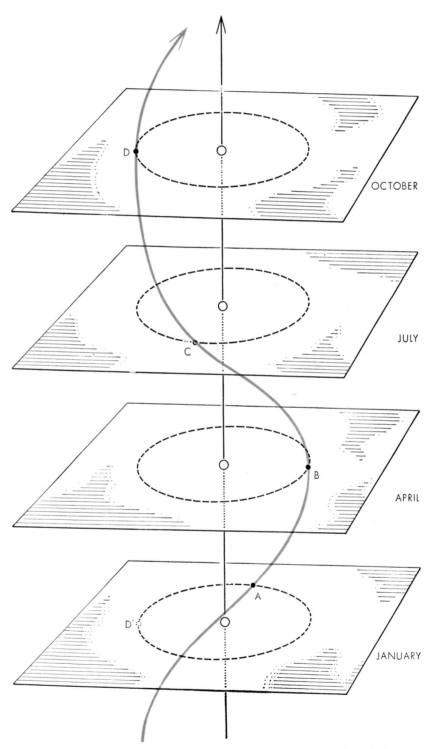

MOTION OF BODY in the curved "space-time continuum" of Albert Einstein is represented by the "world line" of the earth's motion around the sun. Here the sun is the small open circle in each of the four planes. The earth is the black dot on the elliptical orbit. Each plane shows the position of the earth at a month of the year. The world line is in color.

self and became known as "Einstein's spherical universe," gave the space of the cosmos a positive curvature. Like a sphere, this universe was closed and thus had a finite volume. The space coordinates in Einstein's spherical universe were curved in the same way as the latitude or longitude coordinates on the surface of the earth. However, the time axis of the space-time continuum ran quite straight, as in the good old classical physics. This means that no cosmic event would ever recur. The two-dimensional analogy of Einstein's space-time continuum is the surface of a cylinder, with the time axis running parallel to the axis of the cylinder and the space axis perpendicular to it [see drawing at left on this page].

The other static solution based on the mysterious repulsion forces was discovered by the Dutch mathematician Willem de Sitter. In his model of the universe both space and time were curved. Its geometry was similar to that of a globe, with longitude serving as the space coordinate and latitude as time [drawing at right on this page].

Unhappily astronomical observations contradicted both Einstein's and de Sitter's static models of the universe, and they were soon abandoned.

In the year 1922 a major turning point came in the cosmological problem. A Russian mathematician, Alexander A. Friedman (from whom the author of this article learned his relativity), discovered an error in Einstein's proof for a static universe. In carrying out his proof Einstein had divided both sides of an equation by a quantity which, Friedman found, could become zero under certain circumstances. Since division by zero is not permitted in algebraic computations, the possibility of a nonstatic universe could not be excluded under the circumstances in question. Friedman showed

that two nonstatic models were possible. One pictured the universe as expanding with time; the other, contracting.

Einstein quickly recognized the importance of this discovery. In the last edition of his book *The Meaning of Relativity* he wrote: "The mathematician Friedman found a way out of this dilemma. He showed that it is possible, according to the field equations, to have a finite density in the whole (three-dimensional) space, without enlarging these field equations ad hoc." Einstein remarked to me many years ago that the cosmic repulsion idea was the biggest blunder he had made in his entire life.

Almost at the very moment that Friedman was discovering the possibility of an expanding universe by mathematical reasoning, Edwin P. Hubble at the Mount Wilson Observatory on the other side of the world found the first evidence of actual physical expansion through his telescope. He made a compilation of the distances of a number of far galaxies, whose light was shifted toward the red end of the spectrum, and it was soon found that the extent of the shift was in direct proportion to a galaxy's distance from us, as estimated by its faintness. Hubble and others interpreted the red-shift as the Doppler effect—the well-known phenomenon of lengthening of wavelengths from any radiating source that is moving rapidly away (a train whistle, a source of light or whatever). To date there has been no other reasonable explanation of the galaxies' red-shift. If the explanation is correct, it means that the galaxies are all moving away from one another with increasing velocity as they move farther apart.

Thus Friedman and Hubble laid the foundation for the theory of the expanding universe. The theory was soon developed further by a Belgian theoretical astronomer, Georges Lemaître. He proposed that our universe started from a

highly compressed and extremely hot state which he called the "primeval atom." (Modern physicists would prefer the term "primeval nucleus.") As this matter expanded, it gradually thinned out, cooled down and reaggregated in stars and galaxies, giving rise to the highly complex structure of the universe as we know it today.

Until a few years ago the theory of the expanding universe lay under the cloud of a very serious contradiction. The measurements of the speed of flight of the galaxies and their distances from us indicated that the expansion had started about 1.8 billion years ago. On the other hand, measurements of the age of ancient rocks in the earth by the clock of radioactivity (i.e., the decay of uranium to lead) showed that some of the rocks were at least three billion years old; more recent estimates based on other radioactive elements raise the age of the earth's crust to almost five billion years. Clearly a universe 1.8 billion years old could not contain five-billion-year-old rocks! Happily the contradiction has now been disposed of by Walter Baade's recent discovery that the distance yardstick (based on the periods of variable stars) was faulty and that the distances between galaxies are more than twice as great as they were thought to be. This change in distances raises the age of the universe to five billion years or more.

Friedman's solution of Einstein's cosmological equation, as I mentioned, permits two kinds of universe. We can call one the "pulsating" universe. This model says that when the universe has reached a certain maximum permissible expansion, it will begin to contract; that it will shrink until its matter has been compressed to a certain maximum density, possibly that of atomic nuclear material, which is a hundred million million times denser than water; that it will then begin to expand again—and so on through the cycle *ad infinitum*. The other model is a "hyperbolic" one: it suggests that from an infinitely thin state an eternity ago the universe contracted until it reached the maximum density, from which it rebounded to an unlimited expansion which will go on indefinitely in the future.

The question whether our universe is actually "pulsating" or "hyperbolic" should be decidable from the present rate of its expansion. The situation is analogous to the case of a rocket shot from the surface of the earth. If the velocity of the rocket is less than seven miles per second—the "escape velocity"—the rocket will climb only to a certain

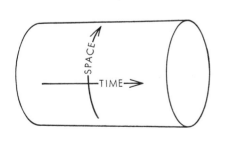
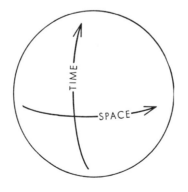

SPHERICAL UNIVERSE of Einstein may be represented in two dimensions by a cylinder (*left*). Its space coordinates were positively curved but its time coordinate was straight. The spherical universe of Willem de Sitter had positively curved coordinates (*right*).

height and then fall back to the earth. (If it were completely elastic, it would bounce up again, etc., etc.) On the other hand, a rocket shot with a velocity of more than seven miles per second will escape from the earth's gravitational field and disappear in space. The case of the receding system of galaxies is very similar to that of an escape rocket, except that instead of just two interacting bodies (the rocket and the earth) we have an unlimited number of them escaping from one another. We find that the galaxies are fleeing from one another at seven times the velocity necessary for mutual escape.

Thus we may conclude that our universe corresponds to the "hyperbolic" model, so that its present expansion will never stop. We must make one reservation. The estimate of the necessary escape velocity is based on the assumption that practically all the mass of the universe is concentrated in galaxies. If intergalactic space contained matter whose total mass was more than seven times that in the galaxies, we would have to reverse our conclusion and decide that the universe is pulsating. There has been no indication so far, however, that any matter exists in intergalactic space, and it could have escaped detection only if it were in the form of pure hydrogen gas, without other gases or dust.

Is the universe finite or infinite? This resolves itself into the question: Is the curvature of space positive or negative—closed like that of a sphere, or open like that of a saddle? We can look for the answer by studying the geometrical properties of its three-dimensional space, just as we examined the properties of figures on two-dimensional surfaces. The most convenient property to investigate astronomically is the relation between the volume of a sphere and its radius.

We saw that, in the two-dimensional case, the area of a circle increases with increasing radius at a faster rate on a negatively curved surface than on a Euclidean or flat surface; and that on a positively curved surface the relative rate of increase is slower. Similarly the increase of volume is faster in negatively curved space, slower in positively curved space. In Euclidean space the volume of a sphere would increase in proportion to the cube, or third power, of the increase in radius. In negatively curved space the volume would increase faster than this; in positively curved space, slower. Thus if we look into space and find that the volume of successively larger spheres, as measured by a count of the galaxies within them, increases

faster than the cube of the distance to the limit of the sphere (the radius), we can conclude that the space of our universe has negative curvature, and therefore is open and infinite. By the same token, if the number of galaxies increases at a rate slower than the cube of the distance, we live in a universe of positive curvature—closed and finite.

Following this idea, Hubble undertook to study the increase in number of galaxies with distance. He estimated the distances of the remote galaxies by their relative faintness: galaxies vary considerably in intrinsic brightness, but over a very large number of galaxies these variations are expected to average out. Hubble's calculations produced the conclusion that the universe is a closed system—a small universe only a few billion light-years in radius!

We know now that the scale he was using was wrong: with the new yardstick the universe would be more than twice as large as he calculated. But there is a more fundamental doubt about his result. The whole method is based on the assumption that the intrinsic brightness of a galaxy remains constant. What if it changes with time? We are seeing the light of the distant galaxies as it was emitted at widely different times in the past—500 million, a billion, two billion years ago. If the stars in the galaxies are burning out, the galaxies must dim as they grow older. A galaxy two billion light-years away cannot be put on the same distance scale with a galaxy 500 million light-years away unless we take into account the fact that we are seeing the nearer galaxy at an older, and less bright, age. The remote galaxy is farther away than a mere comparison of the luminosity of the two would suggest.

When a correction is made for the assumed decline in brightness with age, the more distant galaxies are spread out to farther distances than Hubble assumed. In fact, the calculations of volume are changed so drastically that we may have to reverse the conclusion about the curvature of space. We are not sure, because we do not yet know enough about the evolution of galaxies. But if we find that galaxies wane in intrinsic brightness by only a few per cent in a billion years, we shall have to conclude that space is curved negatively and the universe is infinite.

Actually there is another line of reasoning which supports the side of infinity. Our universe seems to be hyperbolic and ever-expanding. Mathematical solutions of fundamental cosmological equations indicate that such a universe is open and infinite.

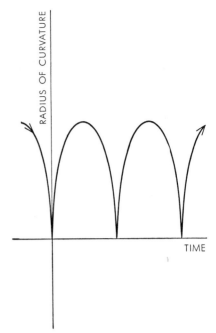

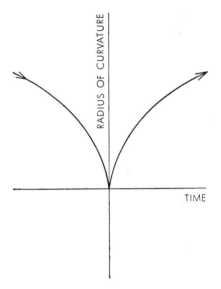

PULSATING AND HYPERBOLIC universes are represented by curves. The pulsating universe at the top repeatedly expands to a maximum permissible density and contracts to a minimum permissible density. The hyperbolic universe at the bottom contracts and then expands indefinitely.

We have reviewed the questions that dominated the thinking of cosmologists during the first half of this century: the conception of a four-dimensional space-time continuum, of curved space, of an expanding universe and of a cosmos which is either finite or infinite. Now we must consider the major present issue in cosmology: Is the universe in truth evolving, or is it in a steady state of equilibrium which has always existed and will go on through eternity? Most cosmologists take the evolutionary view. But in 1951 a group at the University of Cambridge, whose chief spokesman has been Fred Hoyle, advanced the steady-state idea. Essentially their theory is that the universe is infinite in space and time, that it has neither a beginning nor an end, that the density of its matter remains constant, that new matter is steadily being created in space at a rate which exactly compensates for the thinning of matter by expansion, that as a consequence new galaxies are continually being born, and that the galaxies of the universe therefore range in age from mere youngsters to veterans of 5, 10, 20 and more billions of years. In my opinion this theory must be considered very questionable because of the simple fact (apart from other reasons) that the galaxies in our neighborhood all seem to be of the same age as our own Milky Way. But the issue is many-sided and fundamental, and can be settled only by extended study of the universe as far as we can observe it. Hoyle presents the steady-state view in "The Steady-State Universe" [September 1956]. Here I shall summarize the evolutionary theory.

We assume that the universe started from a very dense state of matter. In the

early stages of its expansion, radiant energy was dominant over the mass of matter. We can measure energy and matter on a common scale by means of the well-known equation $E=mc^2$, which says that the energy equivalent of matter is the mass of the matter multiplied by the square of the velocity of light. Energy can be translated into mass, conversely, by dividing the energy quantity by $c^2$. Thus we can speak of the "mass density" of energy. Now at the beginning the mass density of the radiant energy was incomparably greater than the density of the matter in the universe. But in an expanding system the density of radiant energy decreases faster than does the density of matter. The former thins out as the fourth power of the distance of expansion: as the radius of the system doubles, the density of radiant energy drops to one sixteenth. The density of matter declines as the third power; a doubling of the radius means an eightfold increase in volume, or eightfold decrease in density.

Assuming that the universe at the beginning was under absolute rule by radiant energy, we can calculate that the temperature of the universe was 250 million degrees when it was one hour old, dropped to 6,000 degrees (the present

temperature of our sun's surface) when it was 200,000 years old and had fallen to about 100 degrees below the freezing point of water when the universe reached its 250-millionth birthday.

This particular birthday was a crucial one in the life of the universe. It was the point at which the density of ordinary matter became greater than the mass density of radiant energy, because of the more rapid fall of the latter [see chart on this page]. The switch from the reign of radiation to the reign of matter profoundly changed matter's behavior. During the eons of its subjugation to the will of radiant energy (i.e., light), it must have been spread uniformly through space in the form of thin gas. But as soon as matter became gravitationally more important than the radiant energy, it began to acquire a more interesting character. James Jeans, in his classic studies of the physics of such a situation, proved half a century ago that a gravitating gas filling a very large volume is bound to break up into individual "gas balls," the size of which is determined by the density and the temperature of the gas. Thus in the year 250,000,000 A. B. E. (after the beginning of expansion), when matter was freed from the dictatorship of radiant energy, the gas broke up into giant gas clouds, slowly drifting apart as the universe continued to expand. Applying Jeans's mathematical formula for the process to the gas filling the universe at that time, I have found that these primordial balls of gas would have had just about the mass that the galaxies of stars possess today. They were then only "protogalaxies"—cold, dark and chaotic. But their gas soon condensed into stars and formed the galaxies as we see them now.

A central question in this picture of the evolutionary universe is the problem of accounting for the formation of the varied kinds of matter composing it—i.e., the chemical elements. The question is discussed in detail in "The Origin of the Elements" [September 1956]. My belief is that at the start matter was composed simply of protons, neutrons and electrons. After five minutes the universe must have cooled enough to permit the aggregation of protons and neutrons into larger units, from deuterons (one neutron and one proton) up to the heaviest elements. This process must have ended after about 30 minutes, for by that time the temperature of the expanding universe must have dropped below the threshold of thermonuclear reactions among light elements, and the neutrons must have been used up in element-building or been converted to protons.

RELATIVE DENSITY OF MATTER AND RADIATION is reversed during the history of an evolutionary universe. Up to 250 million years (broken vertical line) the mass density of radiation (solid curve) is greater than that of matter (broken curve). After that the density of matter is greater, permitting the formation of huge gas clouds. The gray line is the present.

To many a reader the statement that the present chemical constitution of our universe was decided in half an hour five billion years ago will sound nonsensical. But consider a spot of ground on the atomic proving ground in Nevada where an atomic bomb was exploded three years ago. Within one microsecond the nuclear reactions generated by the bomb produced a variety of fission products. Today, 100 million million microseconds later, the site is still "hot" with the surviving fission products. The ratio of one microsecond to three years is the same as the ratio of half an hour to five billion years! If we can accept a time ratio of this order in the one case, why not in the other?

The late Enrico Fermi and Anthony L. Turkevich at the Institute for Nuclear Studies of the University of Chicago undertook a detailed study of thermonuclear reactions such as must have taken place during the first half hour of the universe's expansion. They concluded that the reactions would have produced about equal amounts of hydrogen and helium, making up 99 per cent of the total material, and about 1 per cent of deuterium. We know that hydrogen and helium do in fact make up about 99 per cent of the matter of the universe. This leaves us with the problem of building the heavier elements. I hold to the opinion that some of them were built by capture of neutrons. However, since the absence of any stable nucleus of atomic weight 5 makes it improbable that the heavier elements could have been produced in the first half hour in the abundances now observed, I would agree that the lion's share of the heavy elements may well have been formed later in the hot interiors of stars.

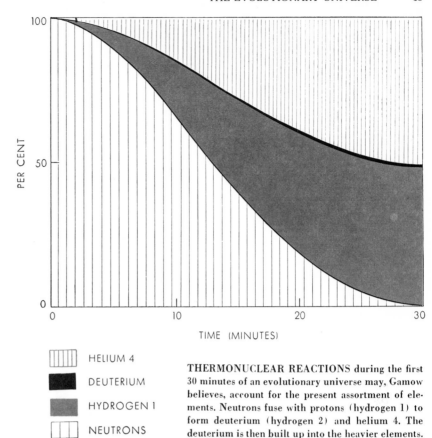

THERMONUCLEAR REACTIONS during the first 30 minutes of an evolutionary universe may, Gamow believes, account for the present assortment of elements. Neutrons fuse with protons (hydrogen 1) to form deuterium (hydrogen 2) and helium 4. The deuterium is then built up into the heavier elements.

All the theories—of the origin, age, extent, composition and nature of the universe—are becoming more and more subject to test by new instruments and new techniques, which are described in later articles in this issue. In the article on the red-shift investigations, Allan Sandage reports a tentative finding that the expansion of the universe may be slowing down. If this is confirmed, it may indicate that we live in a pulsating universe. But we must not forget that the estimate of distances of the galaxies is still founded on the debatable assumption that the brightness of galaxies does not change with time. If galaxies actually diminish in brightness as they age, the calculations cannot be depended upon. Thus the question whether evolution is or is not taking place in the galaxies is of crucial importance at the present stage of our outlook on the universe.

# 3

# The Curvature of Space in a Finite Universe

by J. J. Callahan
*August 1976*

*Curvature of a surface is an intrinsic property
that gives rise to distortion of distances on a map.
The same is true for curvature of space, where the
map is Einstein's general theory of relativity*

Is the universe finite or infinite? According to numerous ancient mythologies, it has a complex structure but is nonetheless finite. That viewpoint developed in Greek philosophy and culminated in the cosmology of Eudoxus and Aristotle: the earth is a ball surrounded by a series of concentric crystalline spheres, the outermost sphere carrying the fixed stars and containing within it the entire material universe. The primary purpose of this cosmology was to explain the motions of the planets and other celestial bodies. Each body was carried around the earth by the rotation of the sphere in which it was embedded. Nevertheless, an integral part of the theory was that the universe was finite.

Aristotle's picture of the world was widely accepted in medieval Europe; it appears, for example, in scholastic philosophy and in Dante Alighieri's *Divine Comedy*. In fact, Dante actually extended Aristotle's picture in a radical and thoroughly modern way. I shall take up Dante's interpretation in my conclusion. In spite of the popularity of the finite-world picture, however, it is open to a devastating objection. In being finite the world must have a limiting boundary, such as Aristotle's outermost sphere. That is impossible, because a boundary can only separate one part of space from another. This objection was put forward by the Greeks, reappeared in the scientific skepticism of the early Renaissance and probably occurs to any schoolchild who thinks about it today. If one accepts the objection, one must conclude that the universe is infinite.

The notion of infinity has always been wrapped in mystery, and historically it triggered apprehensions that have only gradually been overcome. During the scientific Renaissance, Euclidean geometry became the main instrument for comprehending infinite physical space. Euclidean geometry contends that a straight line is the shortest distance between two points, and that the sum of the angles in a triangle will always be 180 degrees. The Renaissance scientists saw that Euclidean geometry treated ideal objects in a mathematical context that was infinite, but that its axioms and propositions exactly described the spatial relations of the real world. Leibniz and Newton shared the view that physical space was infinite and Euclidean. They disagreed, however, on how matter was situated in space. For Leibniz a finite group of stars was unthinkable: such a group would have to be in some specific location in space and God would have had no sufficient reason to put it in one place rather than in some other. Leibniz thus concluded that the universe must be infinite. Newton rejected that possibility, however, on the grounds that God is the only possible actual infinity. Although today these arguments may not seem persuasive, at the time they were considered sufficient.

Who was right? Both arguments were essentially negative. Leibniz denied that the universe was finite, Newton denied that it was infinite. Neither was enthusiastic, however, about the alternative with which he was left. In 1781 Immanuel Kant offered in his *Critique of Pure Reason* a thorough analysis of the entire problem of space, including a bold and novel resolution of the dispute between Newton and Leibniz. Kant said that they were both right and that we must admit paradoxically that the universe is neither finite nor infinite! This basic contradiction between principles that seem equally necessary and reasonable is known as Kant's antinomy of space. The antinomy of space is one of several antinomies that in Kant's view pointed to "a hereditary fault in metaphysics that cannot be explained, much less removed, except by ascending to its birthplace, pure reason itself." A major aim of the *Critique of Pure Reason* was to remove such hereditary faults from metaphysics. Kant's method was drastic. He argued that since we cannot conceive of the universe as being either finite or infinite, we shall never be able to discover empirically whether it is either finite or infinite. In other words, it is not an objective property of the universe to be either finite or infinite. Furthermore, space is not a thing but is a form through which we perceive things, and we make a fundamental error when we treat space as a thing. The antinomy reflected a basic limitation in the mental processes we use to describe the world. Kant would insist that we discard our question as being meaningless.

Today Kant's metaphysical analysis of space is disregarded by modern science because its foundation—notably Euclidean geometry—has been broadened by revolutionary developments in mathematics and physics. Einstein's general theory of relativity provides a new geometry of space, and it opens another approach, unforeseen by Kant, to the question about the finiteness of the world. For Kant the question had simply been invalid. Einstein restored its validity by arguing that Kant's antinomy of space is only apparent and that it can be understood without resorting to metaphysics. In short, Einstein shows that a finite universe is a real possibility.

Like any other physical theory, the general theory of relativity deals with matter and its properties. It regards a galaxy as being perhaps the most natural unit of matter on the cosmic scale. Thus at this level the problem of space is the problem of understanding how the galaxies fit together. A convenient way of visualizing Einstein's solution to the problem is to construct a laboratory model of the entire galactic system. One could construct the model out of balls and sticks, like a model of a large molecule, except that each ball would represent a galaxy and each stick the distance between two galaxies. Before examining Einstein's model, however, let me go back and translate the views of Newton, Leibniz and Kant into the language of models. That may demonstrate more clearly both what they said and how the general theory of relativity went beyond them.

Let us turn first to Newton and his assertion that the universe is finite. If Newton is right, then there can be only a finite number of galaxies. Furthermore, it is reasonable to expect that in time one might devise instruments for locating all of them, and then a complete model of the galactic system could be built. In any case such a model is possible as a mental construct, and that is sufficient for our purposes. If Newton had thought

in terms of such a model, he certainly would have had in mind an exact scale model of the universe: one in which the distances between balls were exactly proportional to the distances between the galaxies they represented.

The main consequence of exact scaling is that any metric feature of the model (that is, any feature that depends only on distance) will be shared by the galactic system. In other words, the model and the galactic system should have the same metric features. The laws of Euclidean geometry are known through direct observation to hold in the terrestrial laboratory, and they dictate all the metric properties of the model. Hence those same laws must dictate the metric properties of the galactic system. That conclusion is very important; in

fact, it is the key to everything that follows. It states that the laws of Euclidean geometry are valid in the galactic system not because they are directly verifiable through observation and measurement in the galactic system but because the system can be reproduced in a scale model. The converse will also be true: if it is impossible to reproduce the galactic system in an exact scale model, then we must abandon the conviction that the geometry of intergalactic space is Euclidean.

Now, a geometric figure and any scale model of it are similar, meaning that corresponding angles in the figure and in the model are identical and corresponding sides are directly proportional to each other in length. Thus we are essentially saying that space can contain simi-

**SPACE WAS FINITE** and had a definite edge, according to the Aristotelian cosmology accepted during medieval times. Here a man is shown looking beyond the edge of space to the Empyrean abode of God beyond. The illustration is often said to be a 16th-century German woodcut; according to Owen Gingerich of Harvard University, it is more likely a piece of art nouveau that was apparently published for the first time in 1907 in *Weltall und Menschheit*, edited by

Hans Kraemer. In either case the picture clearly demonstrates a dilemma posed by Immanuel Kant known as Kant's antinomy of space. Kant believed that the universe had to be finite in extent and homogeneous in composition, and that space had to obey the laws of Euclidean geometry. Actually, however, all those assumptions cannot be true at once. Newton, Leibniz and Einstein had different ways of resolving the dilemma, shown in illustrations on next two pages.

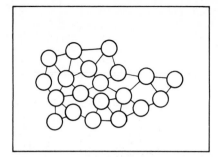 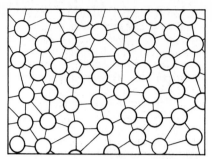

**TWO "BALL AND STICK" COSMOLOGICAL MODELS** demonstrate the philosophical views of Newton and Leibniz; each ball represents a galaxy and each stick represents the distance between galaxies. Although both men accepted Kant's assumption that space obeyed Euclidean geometry, Newton believed the galactic system was finite and inhomogeneous (*left*); thus his model of the galactic system had both a center and a boundary. Leibniz, however, believed the galactic system was infinite and homogeneous (*right*), with no center or boundary.

lar figures of arbitrary sizes if and only if its structure is given by Euclidean geometry. That result is actually very old: it was first obtained by the English mathematician John Wallis in 1663. In either case, using the language of models or of similar figures, we end up with a criterion for determining the geometric properties of space.

Any model of a finite universe has two metric features of special interest. First, the model has a "geographic" center. Second, it has a boundary, consisting of those balls with neighbors on only one side. Therefore if Newton is right, the galactic system must also have a center and a boundary, because it possesses all the metric properties of its scale model.

It is the inhomogeneity of Newton's universe (the fact that not all galaxies have neighbors on all sides) and not its finiteness as such that Leibniz could not accept. Any finite model has a boundary

and a center; in order to eliminate those features one would have to add an infinite number of new balls. Then the model becomes impossible to build. Nevertheless, it is possible to imagine an arrangement of balls reproducing exactly the arrangement of galaxies and fading into the distance in all directions. Let us take that mental construct as Leibniz' scale model. It resembles a model one can build of a crystal, which is also infinite in a theoretical sense. Hence Leibniz' model is actually no stranger than Newton's. If the galaxies are more or less uniformly distributed, then the model will have neither a center nor a boundary.

Like Leibniz, Newton and everyone else in the 18th century, Kant believed in the validity of Euclidean geometry. Unlike many, he knew that Euclidean geometry could not be justified by experience alone. In fact, our key argument,

that we judge space to be Euclidean not through observation but by constructing a model of the galactic system, is due to Kant. In the *Critique of Pure Reason* he actually did not address either models or galaxies. He declared that a direct intuition of space—what I am calling a model—is given to each of us, and through it we discover the properties of space. Because the intuition is universally shared among human beings, it is one of the dictates of pure reason and must be put on an equal footing with sensory experience in investigations of the world. What is more, since the intuition does not depend on experience, which can be faulty or incomplete, the knowledge intuition gives must necessarily be true. As Kant put it, Euclidean geometry is synthetic a priori, by which he meant that it is a special kind of knowledge that is truly descriptive of the world of experience but is not itself derived from that experience. Just as today every science strives to become exact, in the 18th century classical physics saw geometry as its ideal. Kant made his beliefs explicit because he wanted to exploit the special status of geometry to refute the claim of empiricists, most successfully advanced by David Hume, that *all* knowledge of the world is sensory. But is such a direct intuition absolutely necessary to understanding the universe? Must the world admit a scale model? Something fundamental in Kant's philosophy would be undermined if a different situation were to be perceived.

Newton's and Leibniz' models are two clearly distinct models of the galactic system. Kant rejected them both. He had to, because each lacked what he felt was an essential property. For his part he maintained that any study of the material universe must begin by acknowledging three facts: first, the galactic system is finite; second, it is homogeneous and unbounded; third, it can be reproduced in an exact scale model. Those three facts, however, cannot be simultaneously true. In other words, no exact scale model can be both finite and homogeneous. Once again we have arrived at Kant's antinomy of space, this time through the language of models.

Why does the antinomy arise? Kant blamed it on inherent limitations in the mental processes we use to describe the world. In this case the mental process is model building. Kant is saying that we are unable to build a model of the galactic system because our minds cannot tell us how.

There is a way to get out of the antinomy: simply refuse to accept one of Kant's "facts." Remember, not one of them is a physical fact established by direct observation. They are all intuitive assumptions. Even Kant admitted that.

**EINSTEIN'S COSMOLOGICAL MODEL** achieves Kant's desideratum of a finite and homogeneous universe, but only by rejecting Kant's assumption that space was Euclidean. To say that space at large is curved means that it may not be possible to build an exact scale model of the galactic system in the laboratory, where the laws of Euclidean geometry rule. For example, five equidistant galaxies may exist in space, but any attempt to construct a scale model of such a system will fail (*left*). Einstein resolves Kant's antinomy of space by suggesting that when one builds the ball-and-stick model of the system, one should just say that one long stick joining two of the galaxies represents same distance in space as the nine shorter sticks (*right*).

scale model of the earth's surface. It is too good, however, because it reproduces all the earth's geometric features, extrinsic as well as intrinsic, with perfect clarity. Ironically, using it denies us the understanding we seek. We need a model that captures all the intrinsic geometry but at the same time filters out the extrinsic. Let us turn now to the other familiar model of the earth: an atlas of maps.

In many ways any atlas is inferior to a globe. It represents the earth as a collage of overlapping flat charts, and each chart distorts distances. Nevertheless, all the intrinsic geometry of the earth's surface can be recovered from an atlas. This is actually a surprising result mathematically, and it is difficult to prove. The fact remains that the inevitable distortions, although they are a nuisance,

are manageable. How else could worldwide air and sea navigation be based on charts? One might object that an atlas does not look like the earth, that it does not help one to grasp the earth's extrinsic geometry. For our purposes, however, the flatness of the charts, far from being a disadvantage, is their greatest virtue. It implies that one can understand all the intrinsic geometry of a sphere without ever leaving a flat two-dimensional plane. That fact is a tremendous economy, and it shows what abstraction can achieve: one does not need a third dimension in order to understand the structure of the weather network; hence one does not need a fourth dimension in order to understand the structure of the galactic system. Furthermore, what seems to be a creaky analogy between the earth and space

is in fact a real analogy between their abstract intrinsic structures: Einstein's model (the general theory of relativity) is an atlas of space.

How does curvature fit in? The way it is used in the general theory of relativity can be traced back to the work of Carl Friedrich Gauss, in a theory of curved surfaces he published in 1827. Gauss was the first to recognize that a surface has a separate intrinsic geometry. His most remarkable discovery, however (and he even called it that), was that the curvature of a surface is an intrinsic property. Basically Gauss said the following: Take a small portion of a curved surface and flatten it out on a plane. In other words, make a chart. This can generally be done only by stretching the surface, that is, by distorting distances. It should be evident that the original cur-

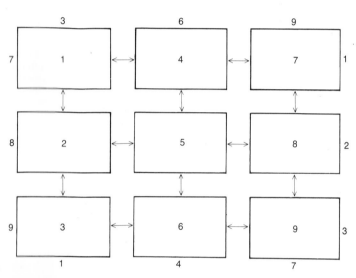

**AN ATLAS FOR A TORUS** (*left*) might consist of nine charts; the numbers around the edges of each chart indicate which of the other

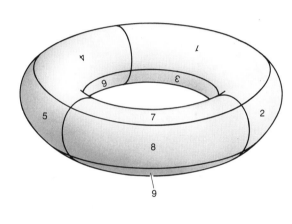

charts it overlaps. The numbered regions on the torus corresponding to the nine different charts in the atlas are shown at the right.

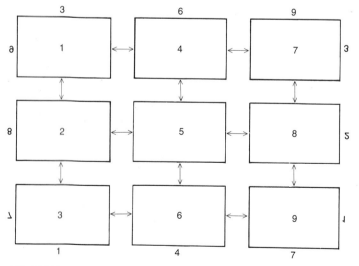

**AN ATLAS FOR A KLEIN BOTTLE** (*left*) might also consist of nine charts. It is very similar to the atlas for the torus except in the way in which the charts overlap. That change, however, makes it im-

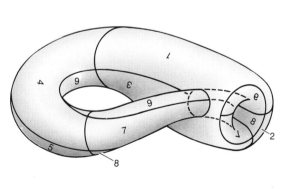

possible to paste the charts together without making the surface intersect itself in places where it should not. The Klein bottle is a form that has one surface. It has no "inside," as a torus and a sphere do.

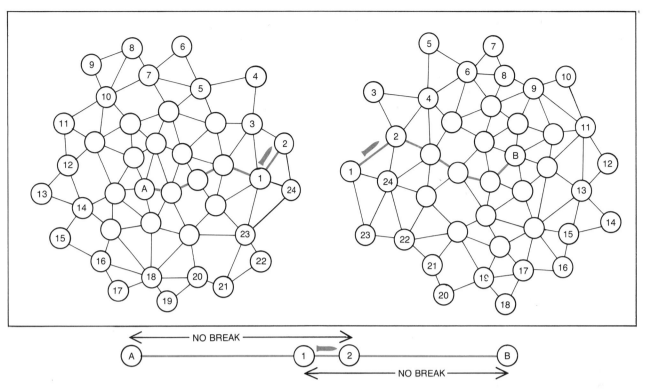

**EINSTEIN'S MODEL** for a finite galactic system is shown here in somewhat more detail. The model is entirely contained within the limits of the illustration. The model appears to be highly inhomogeneous: it is shown in two disconnected pieces each with a center and a boundary, and not all galaxies appear to have neighbors on all sides. Furthermore, two numbered balls, one in each piece, represent the same galaxy. Unnumbered balls represent distinct galaxies that are not duplicated. The model is not built to scale, however; therefore its peculiar properties are not necessarily reflected in the galactic system itself. In fact, the galactic system is actually quite homogeneous. The color line between ball *A* and ball *B* represents a continuous path through the galactic system being taken by a rocket (also shown twice). The continuous nature of the rocket's path is illustrated at bottom. One can think of the model as a pair of three-dimensional viewing screens in which every galaxy appears on at least one screen, and any galaxy at the edge of one screen also appears on the other.

One possibility is to follow Newton and reject Kant's second assumption that the galactic system is homogeneous and unbounded. Then there is no problem in accepting the remaining two assumptions and building a suitable model. That may in fact be the correct choice, since a finite but inhomogeneous galactic system has never been ruled out experimentally. In any case the antinomy is gone. Another possibility is to follow Leibniz and reject the first assumption, that the galactic system is finite. This choice also removes the antinomy, since a homogeneous but infinite system has not been ruled out experimentally either. Or, finally, we can follow Einstein.

Einstein actually achieved Kant's ambition by constructing a down-to-earth model of a finite homogeneous galactic system. He did it by rejecting Kant's third assumption, that the model must be exactly scaled. For example, in Einstein's model a three-inch stick in one location may represent an intergalactic distance of, say, 50 million light-years, whereas in another location it may represent 60 million light-years. To see the impact of Einstein's innovation consider the following imaginary scene.

One day at some time in the distant future the members of an intergalactic

surveying team return to their home base, having measured all the distances between five galaxies in which they are particularly interested. They exchange information and discover that each of the five galaxies is equidistant from the other four. Immediately they can stick four balls together at the vertexes of a regular tetrahedron to represent four of the equidistant galaxies. Where should the fifth ball go? They can attach it to any three of the other four balls and form a second tetrahedron. It is then equidistant from the three balls but not from the fourth ball. Actually there is no way to build a laboratory model consisting of five equidistant balls. Forget about a grand design for the entire galactic system; here is a mere handful of galaxies presenting a crisis in intuition. As a spokesman for the Euclidean position, Kant would have argued that the surveyors are mistaken, because space "is not like that." But the surveyors have already rechecked their work and there is no mistake. Space *is* like that. The assumption that we can build a scale model of any physical system, which is equivalent to the assumption that the geometry of space is Euclidean, is thus revealed to be an attempt to make reality conform to our preconceptions. Ein-

stein turns that around and makes the model conform to reality: he takes the model already built and declares that one long stick should represent the same intergalactic distance as the other nine joining all the balls.

It is only a short step from five galaxies to Einstein's model of the entire galactic system. The illustration above presents a simplified version of Einstein's model using several dozen balls instead of the millions that might actually be required. The number of balls does not really matter, since a larger model will exhibit the same essential features. Two features of the model are particularly noticeable. First, in this illustration there are two disconnected pieces, each with a boundary; second, sometimes two different balls, one in each piece, represent the same galaxy. Because the model is not built exactly to scale, however, those peculiar properties need not carry over to the galactic system itself. In fact, the system is actually quite homogeneous, as one can see by thinking of the model as a pair of three-dimensional viewing screens. In that case every galaxy appears on at least one of the screens, and to guarantee that none is overlooked, any galaxy appearing at the edge of one screen also appears at the edge of the other screen. That explains why two balls sometimes represent the same galaxy.

Now look at one of the galaxies that is visible in both screens. Half of its neighbors appear in one screen and half in the other. Thus although each ball representing that one galaxy is on the boundary of its own screen, the galaxy itself must be completely surrounded by neighboring galaxies. Since the remaining galaxies appear in the middle of either one screen or the other, they also have neighbors on all sides. Hence the galactic system has no boundary. Furthermore, the galactic system is connected, because an object can move continuously from any one galaxy to any other, although it may be necessary to switch viewing screens at some point in order to follow the motion.

Distance distortion is a third essential feature of the model, but it is not particularly noticeable in such an incomplete illustration. Exactly how much distances are distorted is a technical matter, fully treated in Einstein's detailed model (the general theory of relativity). The simple presence of the distortion is what concerns us, because it reveals a fundamental property of space, that is, of the metric relations of the galactic system. Whether or not a model must distort distances is determined solely by the nature of space. As we have already

seen, if the galactic system admits an exact scale model, space is Euclidean. If no scale model is possible, that fact must likewise be due to some contrary property of space. This property has been given the name curvature.

The word curvature may seem an odd choice for a property of space, and so I shall discuss the reason for it below. In any event curvature of space refers simply to the need for distorted models and has nothing to do with space being somehow mysteriously bent. Taking Einstein's model into consideration, one can now draw one of the fundamental conclusions of the general theory of relativity: Kant's antinomy of space arises from an unwarranted assumption that space is Euclidean. A finite and homogeneous galactic system is conceivable, and it is the curvature of space that makes it so.

The connection between curvature and a finite world is a common subject in popular accounts of relativity, but it is usually explained by drawing an analogy with what can happen on surfaces. On a flat plane any finite set of points has a boundary. On the spherical surface of the earth, however, one can imagine a more or less uniformly distributed network of, say, weather stations, and although there are a finite number of stations, every station has neighbors on all sides, so that the net-

work has no boundary. The analogy suggests that the space of a finite homogeneous galactic system must therefore be like the surface of the earth, only one dimension larger. Now, the earth's surface is only two-dimensional, which means that the position of a point on it is determined by just two numbers, latitude and longitude. The reason the earth's surface has no boundary, we feel, is that it curves around into a third dimension. We are led to infer that a three-dimensional "spherical" space must somehow curve around into a fourth dimension. The analogy collapses because it is hopeless to imagine what the extra spatial dimension looks like; no one has ever seen it.

The analogy is actually a good one, but oddly enough it suffers from too detailed a picture of the earth and not too scant a picture of space. A surface has two kinds of geometric properties, intrinsic and extrinsic. An intrinsic property is one that refers to measurements that can be carried out on the surface itself; any other property is extrinsic. For example, in the third century B.C. Eratosthenes deduced the radius of the earth by combining an intrinsic fact about the earth's surface (the fact that the distance from Syene to Alexandria is 5,000 stadia) with an extrinsic fact (the fact that when the sun is directly overhead at Syene, it is 7.2 degrees from the zenith at Alexandria).

Physical space has no extrinsic geometry that we know of, however, because every spatial property we know relates to figures and measurements made in space itself. Thus space cannot be analogous to the surface of a sphere, because it has nothing comparable to the sphere's extrinsic geometry. An analogy between different objects can exist in our minds only where we find it possible to overlook their differences.

Still, there is a way to conceive of the curvature of space. Our actual goal is to understand what a finite homogeneous galactic system must be like. Here it is useful to consider the example of a network of weather stations on the surface of the earth. Notice that the relevant geometric properties of the network are all intrinsic: each weather station is surrounded by neighbors, and the network of stations, although finite, has no boundary. Since those are intrinsic properties, we lose nothing by ignoring the earth's extrinsic geometry; moreover, since the extrinsic part was the stumbling block all along, we are even better off ignoring it. Irrelevant information can be just as distracting to a mathematician as it is to a reader of mystery stories. In mathematics the job of isolating what is genuinely relevant is called abstraction.

The language of models can help once again. A terrestrial globe is an exact

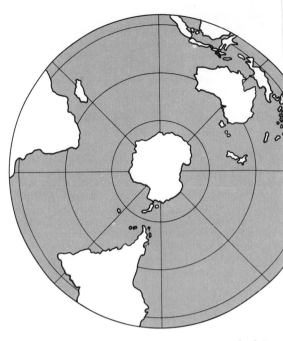

ATLAS OF CHARTS of the spherical earth captures all the intrinsic properties of the earth while filtering out the extrinsic geometry. Since the curved surface of the earth has been flattened out on the charts, distances are distorted. The amount and kind of distortion contain all the information needed to reconstruct the full geo[metry] of the earth. Hence we no longer need a third dimension in or[der to] understand the curvature of the surface of a sphere; similarly, [we do] not need a fourth dimension to understand the curvature of s[pace.]

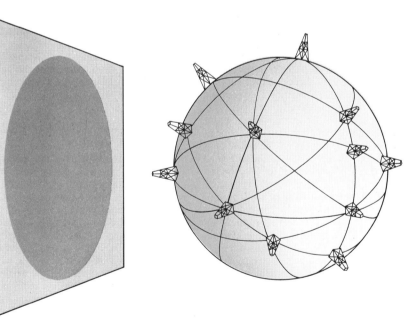

INTRINSIC GEOMETRY AND EXTRINSIC GEOMETRY of the surface of an object such as a sphere are quite different. Any intrinsic property of a surface refers to measurements that can be carried out on the surface itself; any other property is extrinsic. For example, the fact that on a sphere a network of weather stations can be both finite and unbounded is an intrinsic property. The fact that the sphere casts a circular shadow from all viewpoints is an extrinsic geometric feature. Physical space has no known extrinsic geometry because every property we know of relates to figures and measurements made in space itself. Thus it is hopeless to try to imagine curved space as being mysteriously bent through a fourth dimension, since we cannot take an extrinsic view of space by getting outside it and looking back at it.

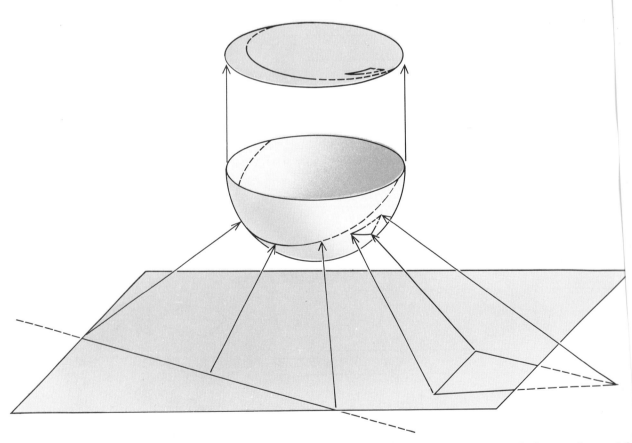

INFINITE PLANE CAN BE MAPPED onto an atlas of charts as well as a sphere can. In this particular case the atlas consists of one chart, the disk, which is made by first projecting the plane onto the hemisphere and then up onto the disk. (This is only one of many ways of mapping the plane.) The geometric shapes again reveal the distance distortions created by the mapping. Any unbounded two-dimensional surface can be mapped by an atlas of charts; such a surface, including the plane and sphere, is a two-dimensional manifold.

vature of the surface determines the precise kind and amount of distortion. Gauss's remarkable discovery is the converse of that fact: the curvature of a surface can be completely determined solely from the distance distortions present in a chart. In other words, distortion of a chart and curvature of a surface are different aspects of the same thing: when the intrinsic geometry of a portion of a surface is abstracted to a chart, its curvature is given by distance distortion. That is why the completely analogous distortions in Einstein's model of the galactic system are attributed to the curvature of space.

Gauss studied the intrinsic geometry of a small portion of a surface by making a chart. In order to map an entire surface it is generally necessary to resort to several charts, or an atlas. A surface that can be mapped by an atlas of charts is called a two-dimensional manifold. The term is meant to emphasize that the surface is usually made up of many two-dimensional patches instead of a single chart.

Every smooth surface without a boundary is a two-dimensional manifold. For example, an infinite plane is a two-dimensional manifold, and so is a torus. One can also start with any arbitrary collection of charts that overlap in a coherent way; they constitute an atlas for some two-dimensional manifold.

The notion of a manifold, which grew out of the attempt to understand the intrinsic geometry of a surface, actually yields something more general. One can make charts out of solid, three-dimensional "blobs" instead of flat, two-dimensional "patches" and have something that makes sense in three dimensions. The resulting object is called, naturally enough, a three-dimensional manifold, and it is a volume rather than a surface. Einstein's model of the universe is an atlas for one particular three-dimensional manifold: it is called the three-dimensional sphere, or three-sphere. (The surface of an ordinary sphere is a two-sphere.) There are countless other examples. The ordinary Euclidean space of mathematics is given by a single chart, analogous to the chart for the infinite plane. Other spaces quickly become too messy to describe in detail. Mathematicians have even defined manifolds for arbitrary numbers of dimensions, by finding a way of getting around the need to visualize the fundamental building blocks: the dimension of a manifold is simply interpreted as the number of variables needed to locate a point on it. Manifolds have thus become a natural setting for problems requiring many variables, and their use now extends beyond mathematics to science in general.

One of the most basic perceptions we have of our environment is that three

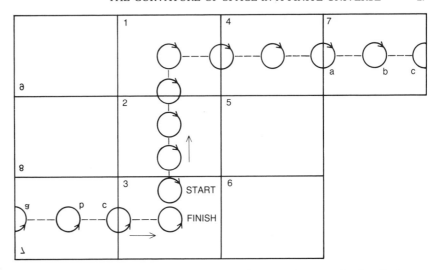

ORIENTABILITY is a global property of a surface. A circle with an arrowhead on its circumference is said to define a local orientation (either clockwise or counterclockwise) when it is drawn on a chart for a manifold. That orientation can be extended to other charts by moving the circle from one chart to another in the atlas. Moving along two different routes, however, may produce conflicting orientations of the circular arrowhead in a distant chart. If that happens, the manifold is said to be globally nonorientable. A Klein bottle is globally nonorientable; shown is its atlas with one chart (*chart 7*) and three of the positions of the circular arrowhead (*a–c*) included twice. When the circular arrowhead returns to its starting position in chart 3, it is an upside-down mirror image of its former self. In some manifolds, for example the torus, such a conflict can never arise; therefore such a manifold is said to be globally orientable.

variables (height, width and breadth, or *x, y* and *z*) are needed to label completely all positions within it. Physical space is a three-dimensional manifold. The mistake of common sense was to assume that it must be the one given by a single chart: Euclidean space. We can now see why this was such a natural error to fall into. Locally (that is, in the immediate neighborhood of any point) all manifolds of the same dimension look alike. In principle they are distinguishable because of the presence of curvature; in practice, however, curvature may not be detectable with sufficient experimental accuracy when it is measured over a small region. (The surveyors in my little scene had to travel to other galaxies to get their results; our own measuring instruments have not yet left the solar system.) In other words, for all practical purposes space *is* Euclidean if it is taken in small enough pieces. The entire cosmos is another matter, and the notion of the manifold provides a wealth of new possibilities for understanding the structure of the universe in the large. Far from being bankrupted by the apparent nonsense of questions about the structure of space, as Kant suggested, mathematics has been substantially enriched.

The most striking differences between one manifold and another are global differences, differences that can be discerned only by studying entire atlases, not single charts. One global property of a surface is orientability. For example, the atlases for a torus and for the surface known as a Klein bottle are quite similar, but the Klein bottle has two peculiar

features that set it quite apart from the torus. First, the Klein bottle cannot be constructed in space without intersecting itself in places where it should not intersect itself. Second, it cannot be oriented in space. Self-intersection cannot be inferred directly from the atlas. It is an extrinsic feature. Nonorientability is an intrinsic feature, however, and it can be discovered by following a moving clock face from one chart in the atlas to the next. Nonorientability is a global property. A second global property is connectivity, arising from the question of whether every closed loop sliced on a surface separates the surface into two pieces. A sphere and a torus differ in their connectivity. There are analogous notions of connectivity for higher-dimensional manifolds. The study of global properties is a part of topology. It thus becomes evident that in a manifold—including Einstein's three-sphere model of space—a number of small localities having a quite ordinary and familiar geometry can be combined to produce a global effect that is both novel and surprising. This aspect of manifolds is also a part of the fascination of the drawings of Maurits C. Escher.

In the search for the origins of Einstein's ideas much attention has been given to the non-Euclidean geometry that was developed in the early 1800's. That emphasis is somewhat misleading. Several mathematicians, including Johann Heinrich Lambert, who lived between 1728 and 1777 and who was a friend of Kant's, had come to the conclusion that a different collection of theorems, logically as sound as those of Euclidean geometry, could be derived from a new system of axioms that differed slightly from Euclid's. A smaller number of 19th-century mathematicians, notably Gauss, János Bolyai and Nikolai Lobachevski, saw further that such a new system could plausibly describe the structure of physical space just as well as Euclid's. It was no longer possible to maintain Kant's position that Euclidean geometry was synthetic a priori. The antinomy of space persisted, however, because the new Lobachevskian space (as it is often called) had the same overall topological structure as Euclidean space. In particular any finite collection of galaxies in it would still have to have a boundary.

The 19th century was a time of tremendous developments in geometry. By the 1870's comprehensive systems embracing all Euclidean, non-Euclidean and projective geometry had evolved. Curiously, the question of whether the geometric systems were physically relevant drifted from the center of attention and even became rather confused. Henri Poincaré, one of the most eminent mathematicians of the time, even maintained that there was no strictly correct geometry for the description of space. The choice of one geometry instead of another was purely a matter of convention, he said, or was no more consequential than the choice between Fahrenheit

and centigrade thermometers to measure temperature.

The mathematics of Einstein's general theory of relativity is quite distinct in style from that of the prevailing systematic geometries at the beginning of the 20th century. It is a part of differential geometry, which, as the name suggests, exploits the power of calculus. With it Einstein showed how to interpret gravitation as being a curvature of space. In other words, distance distortions will be present in a model of even a small portion of space if the space contains an appreciable amount of matter. That distortion, and not gravitational "force," then dictates the paths of moving bodies. Curvature is therefore a local phenomenon as well as a global one. In fact, the general theory of relativity is mainly occupied with the local consequences of curvature. It is difficult to imagine how the geometry of the general theory of relativity, inextricably bound up with matter as it is, could have evolved from the tidy axiomatic systems of Euclid and Lobachevski. Borrowing a locution from the world of the New York theater, where plays can be presented off-off-Broadway, we could say that the geometry Einstein used is non-non-Euclidean. Einstein's ideas were cast in a language very different from even non-Euclidean geometry, called the absolute differential calculus. Until Einstein used it and changed its name to tensor analysis, it had the reputation of being the kind of

pure mathematics that had no connection with the real world.

Ironically tensor analysis did not start that way. Its origins can be found in the work of Bernhard Riemann (1826–1866). On the occasion of Riemann's appointment to the faculty of the University of Göttingen in 1854 at the age of 27 he opened an entirely new line of thinking in his probationary lecture, titled *On the Hypotheses That Lie at the Foundations of Geometry*. The topic had been selected for him by his teacher and colleague, Gauss. The entire modern viewpoint can be found in Riemann's lecture: the concept of an *n*-dimensional manifold; the study of a manifold's intrinsic geometry—particularly curvature—by extending Gauss's work on surfaces; even the radical notion that geometry and physics are inseparable, that is, that the presence of matter determines the curvature of space. Riemann's central concern, however, was with the implications his ideas had for the structure of physical space. His own words (translated) explain it well:

"That space is an unbounded three-dimensional manifold is an assumption that is employed for every apprehension of the external world, by which at every moment the domain of actual perceptions is filled in and the possible locations of a sought-for object are constructed; in these applications it is continually confirmed. For that reason the unboundedness of space has a greater

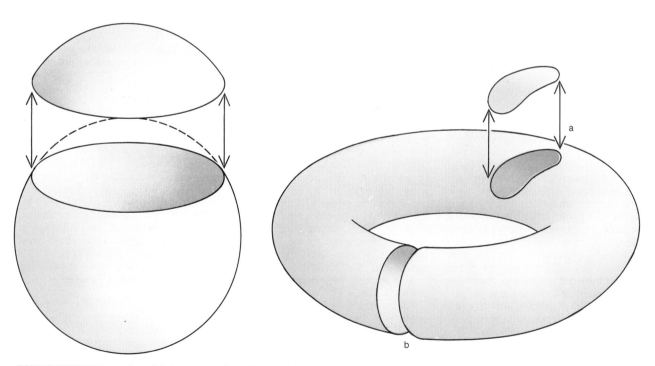

**CONNECTIVITY** is another global property of a surface, and it refers to the question of whether or not every closed loop on a surface separates the surface into two pieces. For example, the connectivity of a sphere and that of a torus are different. Any loop on a sphere separates it into two pieces (*left*). Such loops (*a*) also exist on the torus (*right*), but there are certain loops (*b*) that do not separate it. The fact that nonseparating loops can exist for some surfaces and not for others reflects a basic topological difference between those surfaces.

certainty than any external experience. But its infinitude in no way follows from this. On the contrary, space would necessarily be finite if we assumed that bodies exist independently of position—so that we could ascribe constant curvature to space—as long as this curvature had a positive value, however small."

Riemann treated space as a manifold, and he recognized that one of its possible structures is given by what we have been calling Einstein's three-sphere. Thus it was Riemann who first saw that a finite unbounded universe was conceptually possible. He, not Einstein, actually provided the way around Kant's antinomy of space.

Although Gauss, for one, was astonished by what he heard Riemann say, the wider scientific community did not become generally aware of Riemann's lecture until it was published posthumously in 1868. (Riemann died of tuberculosis at 39.) During the next half-century Riemann's mathematical ideas were developed extensively, but their applications to space were virtually forgotten. A generation later Einstein apparently rediscovered the full wealth of Riemann's ideas about the physical world. Einstein's work, however, is no mere copy of what Riemann had already done. The general theory of relativity is above all a physical theory, a coherent and detailed account of the underlying geometric character of gravitation. Riemann provided the geometric language, but Einstein's physics was radically new. Riemann had only hinted at it.

The positions of Einstein and Kant are by no means antithetical. On the contrary, as we have seen, one of the fundamental tenets of Kant's metaphysical idealism is that space is not a thing but one of the forms through which we organize our perceptions of things. Moreover, the structure of our perceptual organization is given a priori. Those thoughts of Kant's are implicitly accepted by relativity. The basic quarrel between Einstein and Kant is over the structure of space. Kant assumed, partly because he saw no more general possibility, that space must be Euclidean; Einstein maintained that it is Riemannian, and he includes Kant's position as a special case. That explains how the antinomy of space could arise in the first place, and also how Riemann and Einstein could resolve it without completely undermining Kant.

A final word about Dante, following some observations made by Andreas Speiser in his book *Klassische Stücke der Mathematik.* In the first two books of *The Divine Comedy* Dante traverses the material world from the icy core of the earth, the abode of Lucifer, to the Mount of Purgatory. In the last

"WATERFALL," a black-and-white lithograph by Maurits C. Escher, is an example from the world of art of how space might be locally quite ordinary and yet globally quite surprising.

book, *Paradise,* Dante's beloved Beatrice guides him up through the nine heavenly spheres, each sphere larger and more rapidly turning than the last, until he reaches the Primum Mobile, the ninth and largest sphere and the boundary of space. His goal is to see the Empyrean, the abode of God. It finally appears to him, as a blinding point of light surrounded by nine concentric spheres that represent the angelic orders responsible for the motions of the material spheres. Dante is puzzled, however, because the smaller the radius of each Empyrean sphere is, the faster the sphere turns. Beatrice explains that there is no paradox between the material and the spiritual spheres; every sphere, whether material or spiritual, turns faster the more perfect or divine it is. The spiritual world completes the material world exactly as one viewing screen of Einstein's

model of the galactic system completes the other. The overlap between the two is revealed by the correspondence between the celestial spheres and their angelic counterparts, and again as in Einstein's model the farther a sphere is from the center of one chart, the nearer its counterpart is to the center of the other. And the speeds of the material spheres and of the spiritual spheres are in harmony.

Speiser suggests that Dante was able to come to this remarkable vision because his geometric knowledge was derived from astronomy and not from Euclid, which he scarcely knew. Here is a translation by Barbara Reynolds (Penguin, 1962) of the conversation between Dante and Beatrice in Canto XXVIII:

About this Point a fiery circle
    whirled,

With such rapidity it had outraced
  The swiftest sphere revolving round
  the world.

This by another circle was embraced,
  This by a third, which yet a fourth
  enclosed;
  Round this a fifth, round that a
  sixth I traced.

Beyond, the seventh was so wide
  disclosed
  That Iris, to enfold it, were too
  small,
  Her rainbow a full circle being
  supposed.

So too the eighth and ninth; and each
  and all

More slowly turned as they were
  more removed
  Numerically from the integral.

Purest in flame the inmost circle
  proved.
  Being nearest the Pure Spark, or so
  I venture,
  Most clearly with Its truth it is
  engrooved.

Observing wonder in my every feature,
  My Lady told me what I set below:
  "From this Point hang the heavens
  and all nature.

Behold the circle nearest it and know
  It owes its rapid movement to the
  spur

Of burning love which keeps it
  whirling so."

"If manifested in these circles were
  The cosmic order of the universe,
  I should be well content," I
  answered her;

"But in the world below it's the
  reverse,
  Each sphere with God's own love
  being more instilled
  The further from its centre it
  appears;

Whence, if my longing is to be
  fulfilled,
  Here in this wondrous and angelic
  fane,
  Where love and light alone the
  confines build,

I must entreat thee further to explain
  Why copy from its pattern goes
  awry,
  For on my own I ponder it in vain."

"There's naught to marvel at, if to
  untie
  This tangled knot thy fingers are
  unfit,
  So tight 'tis grown for lack of will
  to try."

Then she went on: "This is no meagre bit
  I'll give to thee. Wouldst thou be
  filled? Then take,
  And round its content ply thy subtle
  wit.

Material circles in the heavens make
  Their courses, wide or small, as
  more or less,
  Through all their parts, of virtue
  they partake.

The greater good makes greater
  blessedness;
  More blessedness more matter
  must enclose,
  If all its parts have equal
  perfectness.

It follows that the sphere, which as it
  goes
  Turns all the world along, must
  correspond
  To this, the inmost, which most
  loves and knows.

Hence, if thou wilt but cast thy measure
  round
  The angels' *power*, not their
  circumference
  As it appears to thee, it will be
  found

That wondrous is the perfect congru-
  ence
  Which every heaven with every
  mover shows
  Between their corresponding
  measurements."

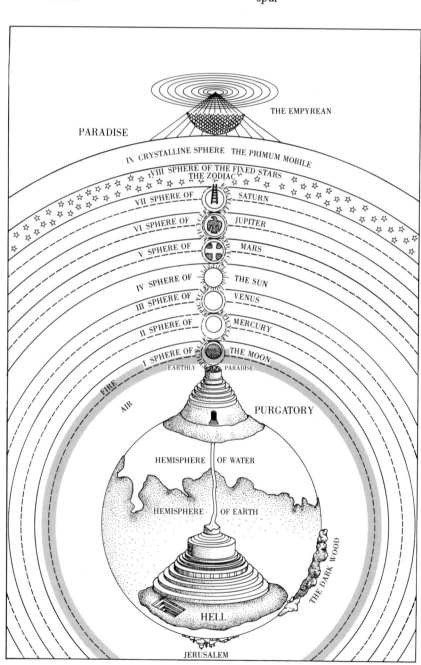

**DANTE ALIGHIERI'S SCHEME OF THE UNIVERSE** in illustration from "Paradise" in *The Divine Comedy* extends Aristotelian cosmology in a modern way. It is discussed in text.

# Cosmology before and after Quasars

by Dennis Sciama
*September 1967*

A review of *The Measure of the Universe*, by J. D. North. Oxford University Press, 1965.

I have often wondered what it must have been like to be a nuclear physicist in the early 1930's, particularly in 1932—that *annus mirabilis* which saw the discovery of the neutron and the positron and the first splitting of the nucleus by artificially accelerated particles. Now I think I know. As a cosmologist I have seen in the 1960's a similar stream of discoveries following one on another at an almost indecent rate. The years 1963 to 1965 stand out, beginning with the discovery of quasars, followed by the measurement of the fantastic red shifts possessed by some of them and culminating in what is perhaps the greatest discovery of them all: the cosmic black-body radiation. I should say at once that the evidence is not yet decisive that any of these discoveries has cosmological significance, but it is good enough to have reduced most cosmologists, who are traditionally starved for basic observations, to a state of bewildered euphoria.

These reflections are prompted by the publication of an interesting book called *The Measure of the Universe,* written by J. D. North, an Oxford philosopher. It is a history of modern cosmology that ends just before the new period begins. It barely mentions quasars and does not mention the cosmic black-body radiation at all. This is no criticism, because the book was written too early for it to have done so. It was therefore written at the right time to take stock of the first great period in cosmology. That period, which had only the expansion of the universe to explain, we might justly call the geometrical period. Today we are well and truly launched into the astrophysical period.

To be fair to the early theorists, they did predict the expansion of the universe before it was discovered. By the early 1920's it was clear to Willem de Sitter, Alexander Friedmann and Hermann Weyl that Einstein's field equations of general relativity had as solutions homogeneous and isotropic model universes whose material substratum was in a state of expansion, the relative velocity at which two particles moved apart simply being proportional to their distance (except for refinements for very widely separated particles). Moreover, if the debatable cosmical constant was dropped from the field equations, as Einstein later urged, then *all* the homogeneous and isotropic solutions exhibited expansion (or contraction if one cared to reverse the sense of time). It was not until 1929 that the Hubble law, that the observed red shift is proportional to the distance of a galaxy (as estimated by various more or less dubious criteria), was first stated.

North's book gives a thorough account of this classical phase of theoretical cosmology. There were many controversies at the time about the properties of the various models. That phase is now over, and the correct results are enshrined in standard theory. There were also controversies of a different nature; not everyone accepted the view that general relativity was uniquely fitted to deal with the universe as a whole. Various "heretical" theories were proposed, notably by Sir Arthur Eddington, E. A. Milne, P. A. M. Dirac and Pascual Jordan, and they are described too. I deliberately mention separately the steady-state theory of Herman Bondi, Thomas Gold and Fred Hoyle, because I think it is fair to say that of all the heretical theories this is the one that has irritated and excited the most people, has provoked the most

good astrophysics and has more or less survived to the present day.

I say "more or less" because one of the consequences of the new turn of events—of cosmology becoming astrophysical—is that if the red shifts of the quasars are cosmological in origin, and if the universe is filled with black-body radiation, then the chances of the steady-state theory surviving are very small indeed. I want to make clear why this is so, and to discuss what further information we can hope to extract from the new results and their likely future extensions. I must add that for me the loss of the steady-state theory has been a cause of great sadness. The steady-state theory has a sweep and beauty that for some unaccountable reason the architect of the universe appears to have overlooked. The universe in fact is a botched job, but I suppose we shall have to make the best of it.

One of the botches is the existence of a singularity, that is, a moment when the density of the universe was infinite. To be more precise, this is what general relativity requires for the homogeneous and isotropic models to which I have referred. It has sometimes been suggested that the singularity would go away as soon as one admitted that the real universe was neither exactly homogeneous nor exactly isotropic; in such circumstances the galaxies would not move quite radially, and so the matter they are made of would not all have emerged from exactly one point in the past. It has recently been shown by Stephen Hawking and others, however, that the orthodox theory of general relativity, without on the one hand a cosmical constant and on the other assumptions of exact symmetry, still requires the physical properties of the universe to have been singular at some time or times.

It was to avoid such an unpleasant singularity (and for other reasons too) that Bondi, Gold and Hoyle proposed in 1948 a deviation from orthodox general relativity that would allow the continual creation of matter at a rate just compensating for the expansion of the universe. The resulting mean density of the universe (and indeed all its other average properties) would then be independent of time, leading to a steady state that would automatically persist forever. It is this magnificent conception we must now reluctantly abandon.

The first evidence against the steady-state theory came from counts of celestial radio sources, conducted notably by Sir Martin Ryle and his colleagues in Britain, but also by B. Y. Mills and J. G. Bolton in Australia and by M. Ceccarelli in Italy. These counts showed that the number of faint radio sources was far too large compared with the number of bright sources to be compatible with the steady-state theory. This evidence has given rise to much controversy, mainly because the majority of sources concerned have not yet been identified optically. Accordingly inferences drawn from these counts have been surrounded by an aura of uncertainty. A straightforward interpretation, stressed by Ryle and studied in detail by William Davidson and by Malcolm Longair, requires that the radio sources exhibit intrinsic evolution. That is to say, the faint sources, which are mostly at great distances and so are now being seen as they were a long time in the past (because of the time their radio waves take to reach us), must have average properties different from the bright sources, which are mostly relatively near and so are being seen almost contemporaneously with ourselves. Such evolution is of course incompatible with a steady-state universe, but it would be expected in a universe evolving from a dense state to a dilute one, a universe to which one can attach the concept of an age.

Needless to say, there have been several implausible attempts to evade Ryle's argument. My own attempt has turned out to be correct, but not in the way I intended. I proposed before the discovery of quasars that the radio sources in Ryle's catalogue consisted of two different populations. One population was to be the radio galaxies that had already been identified optically and were well known. For the second population I proposed the existence of radio stars in our galaxy, whose distribution between bright and faint sources would explain the anomalous counts but

would have nothing to do with cosmology. It has turned out that a second population of starlike (that is, unresolvable) radio sources does exist. Moreover, they are just the ones responsible for the excess of faint sources (as has been shown independently by Philippe Véron and Longair). But these quasi-stellar radio sources, or quasars, have large red shifts and are therefore not the objects I had in mind.

It is true that a few physicists and astronomers (James Terrell, Geoffrey and Margaret Burbidge, Hoyle) hold, with differing degrees of assurance, that these large red shifts are not cosmological in origin, and that the quasars are within, or relatively close to, our galaxy. This would be in tune with my proposal, but I find their arguments unconvincing. If the red shifts have a Doppler origin, that is, if the quasars are receding from us rapidly as a result of a local explosion, the question arises of why we do not see any blue shifts from quasars fleeing from neighboring galaxies toward us. Of course, if quasar emission is a sufficiently rare process, the nearest such galaxies would be too far away for their quasars to be visible, but then why should we be privileged to witness such a rare event so close to us? Clearly this is possible but unlikely.

On the other hand, if the red shifts do not have a Doppler origin but arise, say, from the Einstein gravitational effect, and if the sources are distributed uniformly in space with the ones observed so far quite close to us, then we would not expect the source counts to manifest an excess of faint objects. The relative number of bright and faint sources should be the same as if there were no red shift (that is, the same as for a uniform distribution of stationary sources) and this is not what is observed. We conclude that the red shifts are most probably cosmological in origin. On this basis Martin Rees and I have carried out an analysis of the red shifts of the quasars, and we find again that there are too many faint sources with large red shifts to be compatible with the steady-state theory.

In weighing the significance of what I have said so far it is important to understand how accidental it is that we should be able to observe such large red shifts so easily. Objects with these large red shifts are so distant that the different cosmological theories make substantially different predictions about them, but such objects are visible only because quasars happen to be a hundred times brighter than galaxies. In

contrast, the existence of cosmic black-body radiation, which also serves to distinguish among different theories, is intimately bound up with the development of the universe itself.

The detection in 1965 of excess radiation at microwave frequencies (that is, at wavelengths of a few centimeters) and the evidence that it has a black-body spectrum (that is, is in thermal equilibrium characterized by a single temperature) has been described in "The Primeval Fireball," by P. J. E. Peebles and David T. Wilkinson [see *Scientific American* June 1967]. The temperature observed is about 3 degrees absolute. I should like to make the following comments on this result:

1. No plausible noncosmological explanation has yet been proposed (and not for want of trying).

2. A natural cosmological explanation does exist if the universe was once very dense.

3. A temperature significantly greater than 3 degrees would not be compatible with our general knowledge of radio astronomy and high-energy astrophysics.

I shall say no more about the first point, but I should like to discuss the second and third in a little more detail. As in the case of the expansion of the universe, the existence of cosmic black-body radiation was predicted before it was observed. Around 1950 it was proposed by George Gamow and his associates that the early, dense stages of the universe were very hot, a state of affairs often described as the "hot big bang." Their reason for making this proposal was that in such conditions thermonuclear reactions could occur at an appreciable rate, converting primordial hydrogen into helium and possibly heavier elements. By choosing the right early conditions Gamow was able to account approximately for the abundance of helium with respect to hydrogen that is observed today. This helium problem is actually in a very confused state at the moment, but the important point here is that if the early, dense stages were hot, unquestionably there was ample time for matter and radiation to come into thermal equilibrium. At that time, then, the radiation would have had a black-body spectrum. Moreover, at all times thereafter the spectrum would remain that of a black body, the radiation simply cooling down as the universe expanded. Gamow's original calculations of helium formation led him to predict for the present temperature of the black-body radiation a value of about

30 degrees absolute, but modern calculations are compatible with a lower temperature, in particular with a temperature of 3 degrees absolute.

As I have mentioned in my third point, we now know that a temperature as high as 30 degrees can be ruled out. Cosmic ray protons and electrons interacting with such radiation would produce effects that could be observed, and they are not. Three degrees is about the highest permitted temperature from this point of view, and 3 degrees is just what has been found. The connection between the microwave observations and Gamow's theory was made by Robert H. Dicke and his colleagues at Princeton University. In fact, they had the bad luck to be setting up apparatus to look for the excess radiation when it was discovered accidentally down the road by Arno A. Penzias and Robert W. Wilson of the Bell Telephone Laboratories.

Can the steady-state theory account for the excess radiation? It would be reasonable to propose that along with the newly created matter there comes into existence newly created radiation; indeed, some such effect would be expected as a result of the creation process itself. But why the observed spectrum should be that of a black body over a wide range of wavelengths is totally obscure. It is therefore critically important to establish without doubt that the actual spectrum is that of a black body. The present evidence is strong but not decisive. Further work is being done, and this point should be settled fairly soon.

There is one final property of the radiation that I want to discuss because it is in some ways its most exciting feature, and that is its degree of isotropy—its uniformity with respect to direction of arrival. Just a few months ago R. B. Partridge and Wilkinson announced that any anisotropy is less than a few tenths of a percent. I have heard cynical scientists comment that this result throws doubt on the whole phenomenon, on the grounds that noise generated internally by the observing instruments would be more "isotropic" than externally generated noise, even noise coming from a highly isotropic universe. To silence this cynicism it is necessary to show that the universe is likely to be isotropic to the required degree. A first step in this direction has recently been taken by Charles W. Misner, who has shown that for a certain class of model universes any initial anisotropy would be rapidly removed by a rather exotic form of viscosity involving the pairs of neutrinos that would be excited by the high temperatures then prevailing. Misner's program is to allow the universe to start out as irregularly as it wishes and then to show that all irregularities would be damped out by the action of accepted physical processes, except for those irregularities we actually observe (such as clusters of galaxies).

Another intriguing aspect of the isotropy measurements is that they can be used to determine our "absolute" velocity, that is, our velocity with respect to the distant matter that last effectively scattered the radiation. Because of the Doppler effect such a velocity would reveal itself by leading to a slightly higher temperature for the radiation ahead of us and a slightly lower temperature for the radiation behind us. The present limit on the anisotropy corresponds to a velocity limit of 300 or 400 kilometers per second. One contribution to the expected velocity comes from the sun's known rotation around our galaxy of about 250 kilometers per second. Even this, however, depends on the correctness of Mach's principle [see "Inertia," by Dennis Sciama; SCIENTIFIC AMERICAN, February, 1957]. According to Mach's principle, the local nonrotating frame of reference as determined dynamically coincides with the frame in which distant matter is not rotating. The well-known rotation of our galaxy, with which is associated the galaxy's dynamical flattening, would then be a rotation with respect to distant matter, and therefore to the effective sources of the black-body radiation.

To obtain our net motion relative to these sources, however, we must also allow for the peculiar motion of our galaxy in the local group of galaxies (which has been estimated to be about 100 kilometers per second) and a possible systematic motion of our galaxy in the local supercluster. This supercluster is believed by some astronomers (Vera Rubin, Gérard de Vaucouleurs, K. F. Ogorodnikov) to be a flattened system of galaxies rotating around the Virgo cluster. I have recently rediscussed our possible motion in the supercluster and arrived at a tentative rough estimate for our net motion through the black-body radiation of about 400 kilometers per second in the general direction of the center of our galaxy.

Future observations of the black-body radiation should be able to test this prediction, and in view of the uncertainty surrounding the notion of a local supercluster I would not be at all surprised to find that it is wrong. My point is simply that yet another new range of problems has been opened up for the cosmologist by the existence of the black-body radiation. We have come a long way in a few years from the geometrical considerations described in North's book, and we can rejoice. Cosmology has at last become a science.

# 5

# The Cosmic Background Radiation

by Adrian Webster
*August 1974*

*The space between the galaxies is filled with radiation ranging from radio waves to gamma rays. The radiation has been generated by various processes, some of which are traced to the "big bang"*

No part of the universe is empty. The space between the planets contains the "wind" of ionized gas expelled by the sun and the dust that is seen from the earth as the zodiacal light. The space between the stars contains a variety of materials, ranging from the hydrogen whose emission and absorption at the wavelength of 21 centimeters is studied by radio astronomers to the dust that weakens and reddens the light of distant stars. Even on the largest scale of all, the vast reaches of space between the galaxies are not empty. To be sure, no gas or dust or any other form of matter has been detected there, but it is quite clear that the whole of that space is permeated by a uniform background of electromagnetic radiation. This cosmic background radiation has now been detected across most of the electromagnetic spectrum, from radio waves at a wavelength of 300 meters to gamma rays at a wavelength of $10^{-14}$ meter. It provides a wealth of information on the history of the universe back to its origin in the "big bang."

The cosmic background radiation has been measured only within the past decade, but interest in the subject goes back two and a half centuries. Early in the 18th century Edmund Halley asked: "Why is the sky dark at night?" This apparently naïve question is not easy to answer, because if the universe had the simplest imaginable structure on the largest possible scale, the background radiation of the sky would be intense. Imagine a static infinite universe, that is, a universe of infinite size in which the stars and galaxies are stationary with respect to one another. A line of sight in any direction will ultimately cross the surface of a star, and the sky should appear to be made up of overlapping stellar disks. The apparent brightness of a star's surface is independent of its distance, so

that everywhere the sky should be as bright as the surface of an average star. Since the sun is an average star, the entire sky, day and night, should be about as bright as the surface of the sun. The fact that it is not was later characterized as Olbers' paradox (after the 18th-century German astronomer Heinrich Olbers). The paradox applies not only to starlight but also to all other regions of the electromagnetic spectrum. It indicates that there is something fundamentally wrong with the model of a static infinite universe, but it does not specify what.

Olbers' paradox was resolved in 1929, when Edwin P. Hubble showed that the universe is not static but is uniformly expanding. The galaxies are all receding from one another, and the velocity of recession, as it is perceived on the earth, is directly proportional to the galaxy's distance. The velocity of recession has a strong effect on the light traveling from the galaxy to the earth. First of all, with each passing moment the successive photons (quanta of light) emitted by the stars in the galaxy have farther to go in order to reach the earth, so that their rate of arrival is lower than it would have been if the galaxy had been stationary. Second, the Doppler effect shifts the photons to lower frequencies, so that they have less energy.

Together these two effects weaken the light from the stars in a distant galaxy over and above the dimming due to the galaxy's distance. Both effects become particularly strong when the speed of the galaxy is a substantial fraction of the speed of light, because at that point the special theory of relativity must be taken into account. The result of all these weakening effects is that the energy density of starlight does not reach enormous values and cause the sky to be as bright

as the sun. The same argument applies to photons of all other wavelengths.

At this writing the background radiation has been observed in four regions of the electromagnetic spectrum: the radio region, the microwave region, the X-ray region and the gamma-ray region. The background radiation at each of the different wavelengths provides a different kind of information on the history of the universe, so that I shall deal with each region separately, starting at the long wavelengths and continuing toward the short.

The cosmic background radiation in the radio region is detected between the frequencies of one megahertz (million cycles per second) and 500 MHz, corresponding to the wavelengths between 300 meters and 60 centimeters. This radiation is somewhat difficult to detect and measure because our galaxy is itself a source of radio waves. The background flux of radio power from all directions is dominated by this foreground of galactic radiation. Measurements of the brightness of the radio sky at the frequency of 20 MHz in the direction of the Large Cloud of Magellan have shown, however, that some of the radio power comes from outside our galaxy. Within the Large Magellanic Cloud, which is one of two small galaxies close to ours, is a cloud of ionized gas that absorbs radio waves at the frequency of 20 MHz. C. A. Shain of the Commonwealth Scientific and Industrial Research Organization in Australia has found that in the direction of this cloud of ionized gas, designated 30 Doradus, there is a decrease in the radio brightness of the sky. His observations show clearly that the ionized gas of 30 Doradus is absorbing radiation, so that there must be radiation coming from a distance greater than that of the Large Magellanic Cloud and therefore coming from well outside our galaxy.

To study the background radiation over a wide range of frequencies in the radio region a technique developed by radio astronomers at the University of Cambridge is used. Accurate maps of the radio sky are made at a variety of wavelengths and are compared with one another. The background radiation does not vary appreciably in strength between one direction and another, and so it has the effect of increasing all the measurements at one wavelength on any map by a constant amount; the zero level to which the brightness is referred on each map is established by the background radiation. The way in which the zero level varies from map to map at various wavelengths gives the strength of the background radiation at each wavelength.

This mapping technique has revealed that the intensity, or energy density, of the radio background drops off as the spectral frequency increases. That is, the amount of energy at shorter wavelengths is less than that at longer ones. This type of spectrum is commonly encountered in radio astronomy; it is called a nonthermal spectrum to distinguish it from the thermal radiation of a hot gas, where the relation between intensity and frequency is totally different. The nonthermal radiation is believed to be generated by the synchrotron process, in which high-energy charged particles spiraling along the lines of force in a magnetic field emit energy at radio wavelengths.

The origin of the radio background is

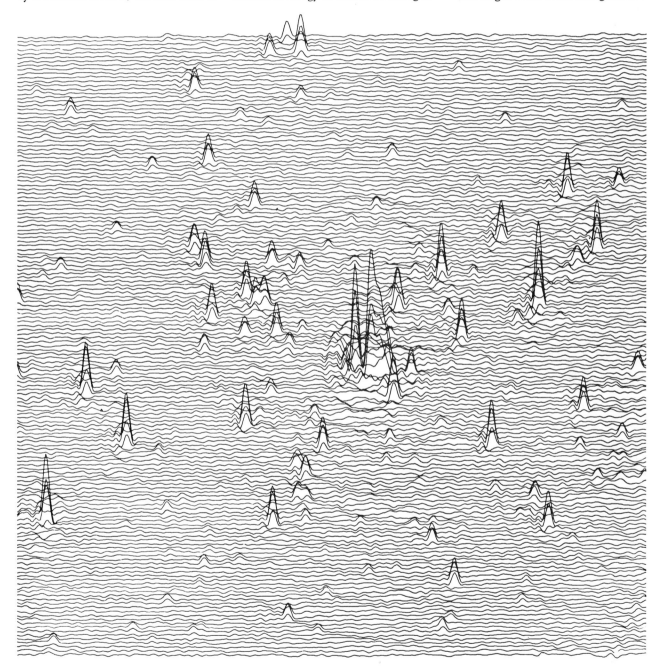

**CHART OF THE RADIO SOURCES** in a small patch of the sky was made by the three dishes of the One Mile Telescope at the University of Cambridge, which are arranged as an interferometer. Each set of peaks on the chart is a radio galaxy or a quasar. Charts such as this one show that there are more faint celestial radio sources than one would expect from the number of bright ones. It is believed that the cosmic background radiation in the radio region of the spectrum is due to all the radio sources in the sky added together. The faint sources are actually powerful sources at great distances, seen now as they were in the distant past because of the time taken for the signals to reach the earth. Such observations indicate sources were more numerous in the past than they are now.

well understood: it is the emission of all the radio sources in the universe added together. These sources, such as radio galaxies and quasars, have nonthermal spectra of just the same kind as the background radiation. It is possible to count enough radio sources on maps made with radio telescopes of the highest sensitivity to infer that the radiation from all of them, including sources too faint to detect individually, adds up to the strength of the radio background. By analyzing the radio background one can study all at once many more radio sources than one could ever hope to detect and examine individually. A great deal of useful information has emerged from such analysis.

For example, the radio background has confirmed evidence from surveys of individual radio sources that there were many more radio galaxies and quasars in the recent cosmic past than there are now. The reason is that if one were to calculate the strength of the background from the present number of radio galaxies and quasars, the result would be much smaller than the observed strength. Allowing for a greater number of these objects in the past makes the calculations agree with the observations. Such confirmation is of vital importance to modern cosmology because it shows, among other things, that the steady-state theory of the universe is quite wrong. The steady-state theory posits that any one

part of the universe is on the average much the same as any other part, and that at present the universe is much the same as it always was and ever will be. One consequence of the steady-state theory is that it calls for the continuous creation of matter to counteract the decrease in density caused by the universe's expansion. The steady-state theory neatly sidesteps all problems about the origin of the universe and the beginning of time by stating that there need not have been a beginning. It is clear, however, that since there were many more radio sources in the past than there are now, the universe was not the same then, and that the fundamental postulate of the steady-state theory is false.

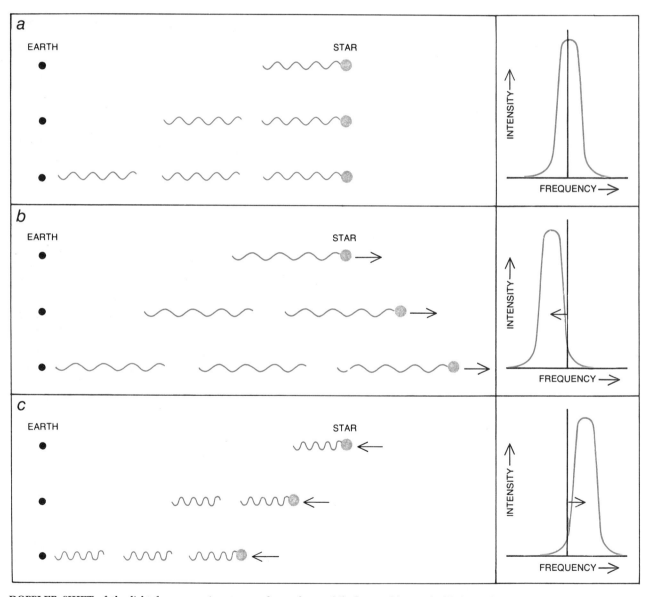

**DOPPLER SHIFT** of the light from a moving star can be used to determine both the velocity and the direction of the star with respect to the earth. If the star is stationary (*a*), its light will be unaffected, and each of the lines in its spectrum will be at the "rest" frequency (*indicated by the vertical line in the spectrum at right*). If the star is moving away from the earth (*b*), its light will be shifted toward longer (redder) wavelengths and will also be dimmed because fewer photons (*wavy line segments*) emitted by the star will reach the earth within a given time interval than if the star were at rest. Conversely, if the star is moving toward the earth (*c*), its light will be shifted toward shorter (bluer) wavelengths and the photons will be arriving at the earth with greater frequency.

The alternative hypothesis to the steady-state theory is the big-bang theory. This theory states that at the beginning of time all the matter in the universe exploded out of a superdense kernel. The observations of the radio background are easy enough to fit into the big-bang theory: the era from about a quarter of the age of the universe up to the present was the era of radio galaxies and quasars. That era is apparently coming to an end, so that there are few of these objects left. In the future there may well be none at all.

One feature of the radio background is conspicuous by its absence. It is well known that in the space between the stars of our galaxy there is a substantial amount of neutral (un-ionized) hydrogen radiating at its characteristic wavelength of 21 centimeters. The 21-centimeter emission has been sought in the spectrum of the cosmic background radiation and has not been found. Consequently there cannot be much neutral hydrogen in the space between the galaxies. This result is of particular interest because according to the canonical hot-big-bang theory, to which I shall return below, the universe must have been full of neutral hydrogen when the initial fireball of the big bang had cooled off. Presumably much of the hydrogen has been used up in the building of stars and galaxies, and the rest may have been re-ionized.

The next region of the spectrum in which the cosmic background radiation has been detected is the microwave region. The microwave background predominates at all frequencies between 500 MHz and 500 GHz (gigahertz, or $10^9$ cycles per second), corresponding to wavelengths between 60 centimeters and .6 millimeter. The microwave background was discovered in 1965 by Arno A. Penzias and Robert W. Wilson of Bell Laboratories [see "The Primeval Fireball," by P. J. E. Peebles and David T. Wilkinson; SCIENTIFIC AMERICAN, June, 1967]. The early measurements of the intensity of the radiation at various wavelengths produced a curve resembling that of a "black body" (an ideal radiator) at a temperature close to 3 degrees Kelvin (degrees Celsius above absolute zero). The radiation seemed to be constant in strength across the sky, and its existence apparently fitted perfectly into the framework of the big-bang theory.

Since that time many more measurements of the microwave background have been made. All have confirmed its black-body curve between 408 MHz and

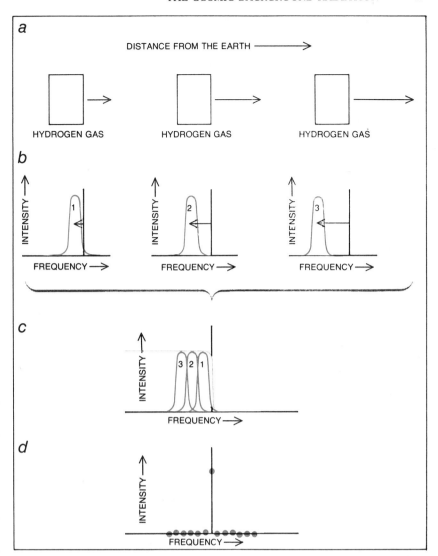

NEUTRAL (UN-IONIZED) HYDROGEN GAS does not fill the space between the galaxies, at least insofar as it has not been detected by radio telescopes. If the universe were filled with hydrogen, presumably the gas that is farther from the earth would be receding faster than the gas that is closer (*a*), just as the galaxies are observed to recede. Therefore the emission line at the radio wavelength of 21 centimeters (corresponding to a frequency of 1,420 megahertz) of the gas farther from the earth (*b*) would be Doppler-shifted more toward longer wavelengths and lower frequencies than the emission line of the gas closer to the earth would be. The result would be that all the shifted emission lines from the gas at the various distances would add up (*c*) to produce a "step" (*light color*) in the radio spectrum below 1,420 MHz. No such step, however, has ever been found (*dots in color*) in the radio spectrum (*d*). The clear signal at exactly 1,420 MHz is from gas in our own galaxy.

115 GHz, corresponding to wavelengths between 74 centimeters and 2.6 millimeters. The best and most recent value for the characteristic temperature of the radiation is 2.76 degrees K.; the measurements are so accurate that it is unlikely that this figure differs from the true one by more than about 3 percent. Instruments carried aloft by balloons and rockets should soon yield accurate measurements in the highest range of microwave frequencies, between 115 GHz and 500 GHz, corresponding to wavelengths between 2.6 millimeters and .6 millimeter. These measurements must be made from above most of the earth's atmo-

sphere because the atmosphere strongly absorbs radiation of millimeter wavelengths. They are very important measurements because they will provide the final check on whether or not the microwave curve is truly a black-body one.

The canonical hot-big-bang theory is an outline that has recently been developed to account for the overall history of the universe. It has been deduced by starting at the present condition of the universe and working backward in time, using all the known laws of physics to calculate what the main processes were at each stage. The details of the calculations are fascinating; it is quite surprising how much can be reliably deduced about what went on so long ago. Here, however, there is room only for a brief summary.

The illustration on page 41 charts the history of the universe from the time when its overall temperature was $10^{12}$ degrees K. Before that time the universe was full of short-lived exotic particles and antiparticles at tremendous density, temperature and pressure. These particles were in equilibrium with the field of radiation, that is, the particles could interact to produce photons and the photons could interact to produce particles. The higher the energy of the photons was, the more massive and peculiar the particles were that each photon could create. By the time the temperature had dropped to $10^{12}$ degrees K., however, the universe was so cool that the only particles whose existence depended on an equilibrium with photons were pairs of positrons and electrons and pairs of neutrinos and antineutrinos. At that stage some protons and neutrons were left, but all the other heavy particles and antiparticles had been annihilated.

Before the temperature had fallen to $10^{11}$ degrees K. the density of matter, although still great by present standards, was low enough for the neutrinos and antineutrinos to cease interacting with the other particles. From that time on these neutrinos and antineutrinos went their separate ways. They are presumably still around, but at present there

does not seem to be any means of detecting them.

When the temperature had dropped to $10^9$ degrees K., the photons did not have enough energy to supply the rest-mass energy equivalent to the mass of the positron-electron pairs (according to the formula $E = mc^2$). Thus the equilibrium between the photons and the pairs of particles was disrupted: positrons and electrons that recombined to produce photons were no longer replaced by the reverse reaction. The positrons were steadily annihilated, leaving a small excess of electrons.

At about the same temperature the protons and neutrons underwent a series of nuclear reactions that resulted in the formation of helium nuclei (composed of either two protons and one neutron or two protons and two neutrons). Most of the helium in the universe today was formed at that time; the number of helium atoms observed in the present universe is an important check on conditions at that stage of the evolution of the universe. No free neutrons survived this stage. The universe at that time was composed of protons and helium nuclei that, together with the electrons, made up an ionized gas in thermal equilibrium with the photons. The enigmatic neutrinos were also still around. Photons in existence at this stage soon interacted with the ionized gas and do not survive to the present day.

When the temperature had dropped below about 5,000 degrees K., the nuclei

and the electrons of the ionized gas recombined to give rise to an un-ionized gas. This gas did not interact further with the photons, so that those photons were undisturbed from that time on. They are what we detect when we observe the microwave background today. Since the time when they were created the Doppler shift caused by the expansion of the universe has weakened the radiation; although it started off at the time of recombination as black-body radiation with a temperature of some 5,000 degrees K. (corresponding to the radiation field at the surface of the sun), it is now black-body radiation at a temperature of only 2.76 degrees K.

The canonical hot-big-bang theory is the framework within which many astrophysicists are currently attempting to understand all the events in the evolution of the universe. In spite of the great leap in understanding brought about by the discovery of the microwave background, many questions remain to be answered, and the answers will need to be fitted into place. Why and how did galaxies form? Why are they the size they are and not larger or smaller? Why do they rotate? Where do their magnetic fields come from? Why and how did quasars form? Why are there fewer quasars now than there were in the past? These questions and many others are all being attacked by trying to find explanatory processes in the universe described by the canonical hot-big-bang model.

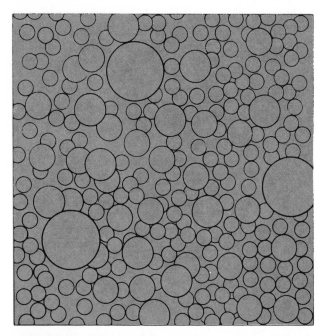

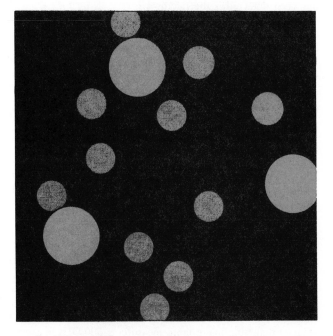

OLBERS' PARADOX states that if the universe were infinite in extent and all the stars were stationary with respect to one another, then the surfaces of the stars could be seen in all directions and the night sky would be very bright (*left*). Since such a phenomenon is not observed, something is wrong with that model of the universe. In fact, it is the expansion of the universe that both reddens and weakens the light of distant stars (*right*) so that the night sky is mostly black. Paradox also applies to other regions of spectrum.

There are some interesting problems at a more fundamental level that seem not so much to invite an answer from within the framework of the model as to demand an explanation in order that the model itself seem less arbitrary. Choosing to work backward from the present state of the universe to gain some knowledge of the initial conditions is not at all arbitrary, but it does not suffice to *explain* the initial conditions. Probably the most we can expect from this approach is that we shall be able at least to *describe* those conditions.

One unanswered question in this category is: Why is the universe expanding at all? We know it is expanding now, and as we work backward we calculate that it was expanding in all the earlier stages as well. But what started the expansion in the first place? There are other such questions, perhaps less obvious but no less important. For example, after the vast numbers of particles and antiparticles had annihilated themselves, why was there a small residuum of real particles (the protons, electrons and neutrons) that constitute the matter in the universe at the present time? Presumably there could just as easily have been a larger or smaller residuum of real particles, or a large or small residuum of antiparticles, or an exact cancellation of the two with no particles or antiparticles left at all. There is nothing in physics that would lead us to prefer any one of these possibilities. Why, then, is there precisely the observed density of real particles? This question goes right back to the initial conditions, because the most likely possibility is that the excess was always there and that it has been a characteristic feature of the universe at all times. It is not yet clear where the answers to the deeper questions are to be found, but these problems are certainly among the most challenging and stimulating in modern astronomy.

Let me return now to the observational situation. With the microwave background, as with the radio background, many attempts have been made to find differences in brightness between different directions in the sky, but no such differences are clearly apparent. The present state of the measurements is that variations in brightness in any angular area of the sky from 180 degrees across to about .1 degree across must be no greater than .1 percent. In other words, the microwave background is very uniform indeed.

It is possible to infer many interesting consequences from this fact. First, the solar system must be moving no faster

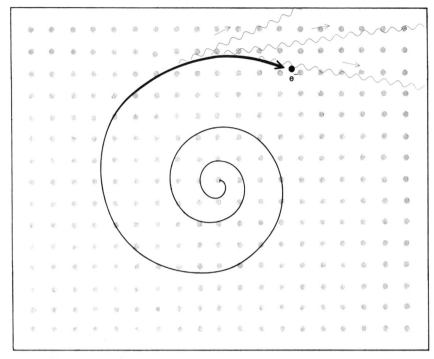

SYNCHROTRON RADIATION (*waves in color*) emitted by a charged particle (in this case an electron) spiraling in a magnetic field probably accounts for background radiation in radio region. Here the lines of magnetic force are perpendicular to the page (*gray dots*). Electron is spiraling up from the page and is emitting radiation tangent to its path.

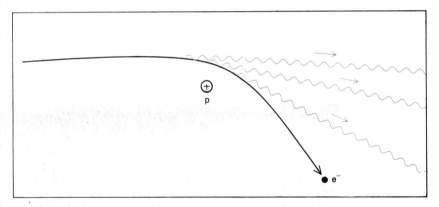

BREMSSTRAHLUNG RADIATION is generated when an electron passes so close to a nucleus (*plus sign in circle*) that its trajectory is bent. The slowing of the electron results in the emission of photons. This mechanism probably gives rise to the cosmic background radiation in the microwave region. (*Bremsstrahlung* is the German for braking radiation.)

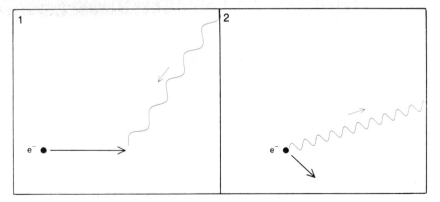

INVERSE COMPTON SCATTERING probably accounts for the background radiation in the X-ray and gamma-ray region. Here a low-energy photon impinges on a high-energy electron (*1*). Electron imparts some of its energy to photon and is itself slowed down (*2*).

than some 300 kilometers per second with respect to the frame of reference defined by the microwave background. If it were, the Doppler effect would cause the radiation from the direction in which the solar system is heading to be brighter than that from the rest of the sky by about .1 percent. It is likely that a small improvement in these observations will reveal the velocity of 200 kilometers per second of the solar system's rotation around the center of our galaxy.

A second consequence is that the universe must be expanding at the same rate in all directions. Otherwise the differences among the Doppler shifts of the radiation from the various regions of the recombining ionized gas that is responsible for the microwave background would cause brightness variations across the sky. A third consequence is that the universe cannot be rotating with any appreciable angular velocity or again the relativistic Doppler effect would give rise to observable brightness variations. The upper limit set for any possible rotation by these measurements is the phenomenally low value of a billionth of a second of arc per year.

The measurements of the uniformity of the background radiation on the smallest angular scales are designed to detect fluctuations in the density of the recom-

bining ionized gas that condensed into the clusters and superclusters of galaxies in the present-day universe. Most calculations predict that the expected nonuniformities in brightness will be very small. Thus it is not too surprising that they have not yet been found.

An intriguing phenomenon within our own galaxy has been discovered through the microwave background. In the interior of certain small, dense clouds of gas and dust along the Milky Way there is a substantial concentration of formaldehyde molecules ($H_2CO$). Radio observations of these clouds have revealed an absorption line in their spectra at a wavelength of six centimeters, a wavelength characteristic of the formaldehyde molecule. At that wavelength the only continuous-spectrum radiation strong enough to be absorbed by the formaldehyde is the microwave background. The fact that the formaldehyde is absorbing energy rather than radiating it indicates that its temperature must be lower than the 2.76 degrees K. of the microwave background. In fact, the temperature of the formaldehyde is about one degree K. This raises the question of how the formaldehyde ever got so cold and how it stays that way. After all, if the formaldehyde were left to itself, it would be warmed to 2.76 degrees by the micro-

wave background radiation. Some kind of previously unsuspected cosmic refrigerator must be working inside the clouds to keep the formaldehyde chilled.

In the X-ray region the cosmic background is represented by radiation at frequencies higher than $2.5 \times 10^{17}$ Hz, corresponding to a wavelength of 1.3 nanometers. This radiation has a nonthermal spectrum that extends well into the gamma-ray region; the highest-energy photons detected so far have an energy of a little more than 100 million electron volts. The origin of the X-ray and gamma-ray background is not yet settled, but it certainly comes from outside our galaxy. The intensity of the radiation is much the same in all directions away from the plane of the Milky Way, where the sources would presumably be concentrated if they were within the galaxy.

The X-ray background is probably the sum of the radiation from a number of discrete sources, just as the radio background is the sum of the radiation from the radio galaxies and quasars. Attempts have been made to detect a graininess in the X-ray background that would reveal whether or not it is coming from a number of discrete sources, but the sensitivity of X-ray telescopes is not yet high enough to reveal the expected fluctuations. Moreover, too few extragalactic X-ray sources have been detected for the contribution of such sources to the X-ray background to be reliably calculated. Nonetheless, it is quite likely that the total emission from all X-ray sources can account for the entire background.

In the gamma-ray region the situation is not so clear. Gamma-ray telescopes are not yet sensitive enough to find any individual sources. The smooth continuity of the spectrum of the background radiation at the X-ray wavelengths and the gamma-ray wavelengths suggests that the gamma rays have the same origin as the X rays, but this need not be true.

The study of the background radiation in the X-ray and gamma-ray regions of the spectrum is still in its infancy. There is surely much interesting information to be gained from such investigations. The full value of even the present observations cannot be realized, however, until the basic question of the radiation's origin is settled. There is no shortage of possible mechanisms for generating the photons; the inverse Compton effect and the *Bremsstrahlung* mechanism are both likely candidates [*see illustrations on page 39*]. Nor is there any lack of hypotheses on the radiation's place of origin. The definitive observa-

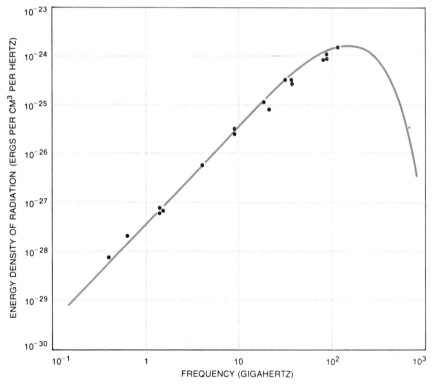

**SPECTRUM OF MICROWAVE BACKGROUND** shows how well the measurements (*dots*) fit the theoretical curve of a black body (an ideal radiator) at a temperature of 2.76 degrees Kelvin. Further measurements are needed in range of frequencies between $10^2$ and $10^3$ gigahertz ($10^9$ cycles per second) to confirm that the spectrum is indeed a black-body one.

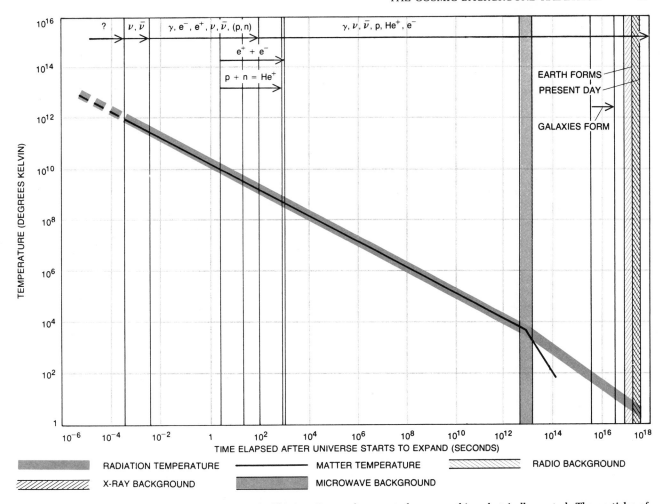

OUTLINE OF EVENTS IN THE UNIVERSE since the "big bang" shows when and how the cosmic background radiation originated in the various regions of the spectrum. Events in the universe within the first one-thousandth ($10^{-3}$) of a second after the big bang are not well understood, principally because they are dominated by interactions of nuclear particles that are not well understood. Up until a hundredth ($10^{-2}$) of a second after the big-bang neutrinos ($\nu$) and antineutrinos ($\bar{\nu}$) are plentiful; they easily interact with photons ($\gamma$) and other particles. Thereafter the neutrinos and antineutrinos do not interact further, and so they play little part in subsequent physical processes. From $10^{-2}$ second until 100 seconds after the big bang the universe consists mostly of photons, electrons ($e^-$), positrons ($e^+$), neutrinos, antineutrinos and a trace of protons ($p$) and neutrons ($n$). During this time and continuing somewhat thereafter all the positrons combine with electrons ($e^+ + e^-$); in addition neutrons combine with protons to make helium nuclei ($He^+$). Most of the helium in the present universe was synthesized at this time. From 100 seconds after the big bang right up to the present the universe consists mostly of photons, neutrinos, antineutrinos, protons, helium nuclei and just enough electrons to keep everything electrically neutral. The particles of the ionized gas of hydrogen and helium interact frequently with the photons of radiation, so that the matter and the radiation stay at the same temperature. That temperature, however, is steadily decreasing as the universe expands. Somewhere in the interval between about 100 and $10^{14}$ seconds after the big bang the energy density of the electromagnetic radiation drops below the rest-mass energy of the matter, so that the photons and the particles no longer interact so freely. The large-scale dynamics of the universe change as a result, and the temperature drops off a little faster than before. At about $10^{13}$ seconds after the big bang electrons recombine with the ionized gas, emitting visible radiation that was subsequently Doppler-shifted by the expansion of the universe to the microwave region of the spectrum; the microwave background is this radiation from the recombination. The gas (matter) and the radiation now cool separately, each at its own pace; the final temperature of the radiation at the present time is 2.76 degrees K. Both the X-ray background and the radio background originate later, when X-ray sources and radio sources came into being. Too little is known about origin of gamma-ray background to include it in illustration.

tions have simply not yet been made.

In the infrared, visible and ultraviolet regions of the spectrum the cosmic background radiation has not yet been detected. There are technical difficulties in the way of some observations in these regions. For example, infrared and ultraviolet observations cannot be made from the ground because the earth's atmosphere strongly absorbs such wavelengths, and a balloon or a rocket is an exceedingly tricky observing platform.

There are natural obstacles in the way of other observations. The faint glow of the earth's atmosphere at night, the zodiacal light and the faint stars in our galaxy are together at least 100 times brighter than the background radiation would be at the visible wavelengths. Direct observations of the background radiation may never be made in the far-ultraviolet region: neutral hydrogen atoms in the interstellar space of our galaxy absorb these photons so strongly that the extra-galactic background radiation probably cannot reach the solar system at all.

As I have indicated, however, the extension of observations of the cosmic background radiation in other regions of the spectrum is far from hopeless. Indeed, it seems certain that we can look forward to observations that will lead to new and exciting inferences about the nature and history of the universe in the large.

# The Evolution of Quasars

6

*May 1971*

*It seems that they were much more plentiful when the
universe was young than they are today. Their light
takes so long to reach us that we can observe millions
of them that have long been extinct*

Since light has a finite velocity the astronomer can never hope to see the universe as it actually exists today. Far from being a handicap, however, the finite velocity of light enables him to peer back in time as far as his instruments and ingenuity can carry him. If he can correctly interpret the complex messages coded in electromagnetic radiation of various wavelengths, he may be able to piece together the evolution of the universe back virtually to the moment of creation. According to prevailing theory, that moment was some 10 billion years ago, when the total mass of the universe exploded out of a small volume, giving rise to the myriad of galaxies, radio galaxies and quasars (starlike objects more luminous than galaxies) whose existence has been slowly revealed during the past half-century.

Optical observations have shed little light on the evolution of ordinary galaxies because even with the most powerful optical telescopes such galaxies cannot be studied in detail if they are much farther away than one or two billion light-years. The astronomer sees them as they looked one or two billion years ago, when they were already perhaps eight or nine billion years old.

Quasars, on the other hand, provide a direct glimpse of the universe as it existed eight or nine billion years ago, only one or two billion years after the "big bang" that presumably started it all.

Some 50 years ago the first large telescopes had shown that the light from distant galaxies is shifted toward the red end of the spectrum; the more distant the galaxy, the greater its red shift and the higher its velocity of recession. Like raisins in an expanding cosmic pudding, all the galaxies are receding from one another. From the observed velocities of recession one can compute that some 10 billion years ago all the matter in the universe was jammed into a tiny volume of space.

The term quasar, a contraction of "quasi-stellar radio source," was originally applied only to the starlike counterparts of certain strong radio sources whose optical spectra exhibit red shifts much larger than those of galaxies. Before long, however, a class of quasi-stellar objects was discovered with large red shifts that have little or no emission at radio wavelengths. "Quasar" is now commonly applied to starlike objects with large red shifts regardless of their radio emissivity.

This article is based on the hypothesis that the quasar red shifts are cosmological, that is, they are a consequence of the expansion of the universe and thus directly related to the distance of the object. On that hypothesis quasars are very remote objects. According to a contrary hypothesis, which will be discussed toward the end of the article, quasars are relatively close objects.

A recent study of quasars carried out with the aid of the 200-inch Hale telescope on Palomar Mountain has provided evidence that these extremely luminous objects evolved quite rapidly when the universe was young. The study indicates that quasars were about 100 times more plentiful when the sun and the earth were formed some five billion years ago than they are today. They were perhaps more than 1,000 times more plentiful at a still earlier epoch, say eight or nine billion years ago. Earlier than that, however, there may have been fewer quasars, perhaps because conditions in the universe had not yet favored their development.

The study embraced all the quasars in two areas of the sky representing a thousandth of the total celestial sphere. By extrapolation one can say with reasonable confidence that a complete sky survey with the largest telescopes should reveal on the order of 15 million quasars. The overwhelming majority are so far away that they almost certainly burned themselves out in the billions of years required for their light to reach us. All of them, of course, can still be studied telescopically, given the time and the inclination. If, however, it were possible to conduct an instantaneous survey of the universe, one might find that only about 35,000 quasars are in existence and radiating with their characteristic intensity at the present time. These find-

BRIGHTEST QUASAR and first member of its class to be recognized is 3C 273, indicated by the reticle in the photograph on the opposite page. The negative print was made from a 1-by-1⅜-inch portion near the edge of a 14-inch square plate taken with the 48-inch Schmidt telescope on Palomar Mountain as part of the National Geographic Society–Palomar Sky Survey. In 1962 the starlike object was found to coincide with the position of a strong radio source designated No. 273 in the third catalogue compiled by radio astronomers at the University of Cambridge. The optical magnitude of 3C 273 is 13. In the entire sky there are at least a million stars of that magnitude. A study of 3C 273's strange spectrum revealed, however, that its light was shifted toward the red end of the spectrum by an amount that indicated the object was receding at about one-sixth the velocity of light. This implied that it was not a nearby star but an object between one billion and two billion light-years away. A galaxy at the same distance would appear at least four magnitudes fainter, which means that 3C 273 is intrinsically at least 40 times brighter. The term "quasi-stellar radio source," or quasar, was coined to describe 3C 273 and other starlike objects exhibiting a large red shift.

ings are in conflict with the "steady state" hypothesis, which holds that the universe has always looked exactly the way it does today. That hypothesis postulates that new matter is continuously being created to maintain the expanding universe at a constant density.

After 10 years of intensive study by optical and radio astronomers quasars remain among the most puzzling of all celestial objects. Assuming that they are at cosmological distances, one can easily show that many quasars are from 50 to 100 times brighter than entire galaxies

containing hundreds of billions of stars. Unlike the light output of normal galaxies, the light output of some quasars has been observed to change significantly in a matter of days. The only explanation is that some variable component of a quasar, if not the entire quasar, may be not much larger than the solar system.

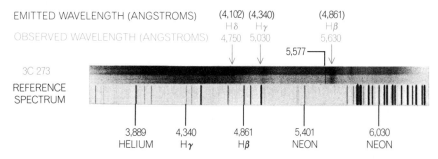

PORTION OF SPECTRUM OF QUASAR 3C 273 shows three prominent emission lines centered at 4,750, 5,030 and 5,630 angstroms, corresponding to the hydrogen emission lines delta, gamma and beta in the Balmer series. The upper and lower halves of the spectrogram were given different exposures to facilitate study. The three emission lines are produced by hydrogen atoms in various states of excitation. Two of the three lines, H gamma and H beta, also appear in the reference spectrum at their normal emission wavelength: 4,340 and 4,861 angstroms. The normal wavelength for H delta is 4,102 angstroms. The red shift, z, is obtained by subtracting the normal wavelength from the observed wavelength and dividing the difference by the normal wavelength. For 3C 273 z is .158, indicating the quasar is receding at nearly a sixth the speed of light. The sharp line at 5,577 often appears in spectra of astronomical objects and serves as a convenient reference point; it is produced by excited oxygen atoms in the upper atmosphere. This spectrogram and others in this article were made by one of the authors (Schmidt), who provided the original interpretation of 3C 273's spectrum.

## The Discovery of Quasars

Before 1960 radio astronomers had identified and catalogued hundreds of radio sources: invisible objects in the universe that emit radiation at radio frequencies. From time to time optical astronomers would succeed in identifying an object—usually a galaxy—whose position coincided with that of the radio source. Thereafter the object was called a radio galaxy. The large majority of radio sources remained unidentified, however, and the general belief was that the source of the emission was a galaxy too far away, or at least too faint, to be recorded on a photographic plate.

In 1960 Thomas Matthews and Allan Sandage first discovered a starlike object at the position given for a radio source in the Third Cambridge ("3C") Catalogue, compiled by Martin Ryle and his colleagues at the University of Cambridge. The radio object 3C 48 coincided in position with a 16th-magnitude star whose spectrum exhibited broad emission lines that could not be identified. Not only did the object emit much more ultraviolet radiation than an ordinary star of the same magnitude but also its brightness varied by more than 40 percent in a year.

Object 3C 48 was thought to be a unique kind of radio-emitting star in our own galaxy until 1963, when the strong radio source 3C 273 was identified with a starlike object of 13th magnitude and one of the authors (Schmidt) recognized that most of the puzzling lines in its spectrum could be explained as the Balmer series of hydrogen lines, shifted in wavelength toward the red by 15.8 percent, or .158 [see illustration on page 43 and upper illustration at left]. Red shifts are commonly expressed as a fraction or percentage obtained by dividing the measured displacement of a line by the wavelength of the undisplaced line. With this clue it was immediately evident that the lines in the spectrum of 3C 48 had a red shift of .367 [see "Quasistellar Radio Sources," by Jesse L. Greenstein; SCIENTIFIC AMERICAN, December, 1963].

Such large red shifts, equivalent to a significant fraction of the velocity of

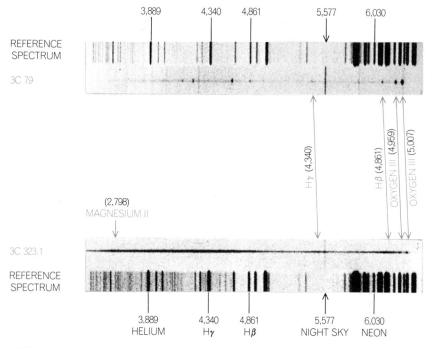

CONTRAST BETWEEN QUASAR AND RADIO GALAXY is shown by these two spectra. The spectrum at the top is that of the strong radio galaxy 3C 79, which has a red shift of .256. The bottom spectrum is that of quasar 3C 323.1, whose red shift is just slightly greater: .264. The radio galaxy produces a substantial number of sharp emission lines. Four of the lines in the right half of its spectrum are identified and compared with their much broadened counterparts as they appear in the spectrum of the quasar. Quasars characteristically emit strongly in the ultraviolet part of the spectrum. A common emitting ion is singly ionized magnesium, designated magnesium II, which has an emission wavelength of 2,798 angstroms.

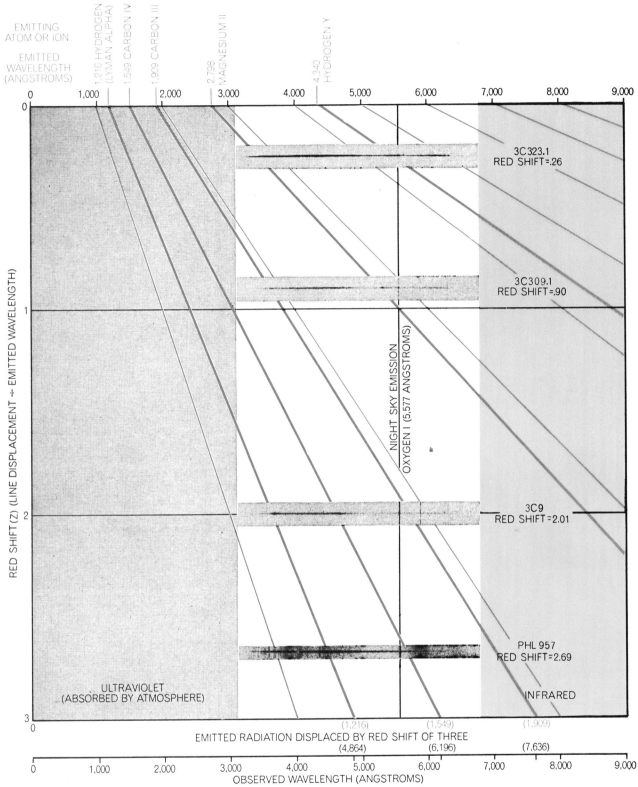

EMITTING
ATOM OR ION

EMITTED
WAVELENGTH
(ANGSTROMS)

1,216 HYDROGEN (LYMAN ALPHA)
1,549 CARBON IV
1,909 CARBON III
2,798 MAGNESIUM II
4,340 HYDROGEN Y

0    1,000    2,000    3,000    4,000    5,000    6,000    7,000    8,000    9,000

0

RED SHIFT (z) (LINE DISPLACEMENT ÷ EMITTED WAVELENGTH)

3C 323.1
RED SHIFT =.26

3C 309.1
RED SHIFT =.90

NIGHT SKY EMISSION
OXYGEN I (5,577 ANGSTROMS)

1

2

3C 9
RED SHIFT =2.01

PHL 957
RED SHIFT =2.69

ULTRAVIOLET
(ABSORBED BY ATMOSPHERE)

INFRARED

3

0

(1,216)    (1,549)    (1,909)
EMITTED RADIATION DISPLACED BY RED SHIFT OF THREE
(4,864)    (6,196)    (7,636)

0    1,000    2,000    3,000    4,000    5,000    6,000    7,000    8,000    9,000
OBSERVED WAVELENGTH (ANGSTROMS)

FOUR QUASAR SPECTRA are positioned on a diagram that shows how radiation emitted at one wavelength billions of years ago is "stretched" on its long journey through space by the presumed expansion of the universe. At least two emission lines are needed to establish the red shift of an astronomical object. A single line could represent any line shifted by any arbitrary amount. Here the heavy slanting lines correspond to the radiation emitted by hydrogen (Lyman alpha), carbon IV, carbon III, magnesium II and hydrogen (gamma). The roman numerals are one greater than the number of electrons missing from the atom. At a red shift, $z$, of 1 the Lyman-alpha line is observed at 2.432 angstroms; at a red shift of 2 the line is observed at 3,648 angstroms; at a red shift of 3 it would appear at 4,864 angstroms. Thus when $z$ equals 2 the initial wavelength is stretched exactly three times; when $z$ equals 3, four times and so on. The quantity $1 + z$ expresses how much the universe has expanded between the emission of a photon and its observation. Only two quasars are known with a red shift greater than 2.5; one of them is PHL 957, whose spectrum appears here. Its spectrum was made with an image-tube spectrograph; the other three spectra were recorded directly on photographic film. The photons that produced the spectrum of PHL 957 left the quasar when the universe was only about 13 percent of its present age.

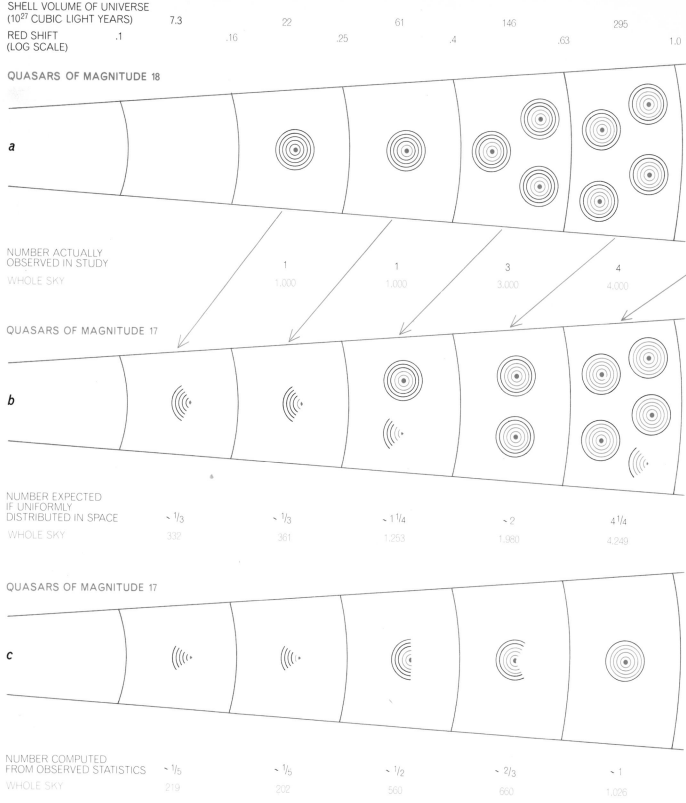

QUASARS OF MAGNITUDE 18

*a*

NUMBER ACTUALLY
OBSERVED IN STUDY                    1              1              3              4
WHOLE SKY                      1.000          1.000          3.000          4.000

QUASARS OF MAGNITUDE 17

*b*

NUMBER EXPECTED
IF UNIFORMLY
DISTRIBUTED IN SPACE      ~1/3          ~1/3          ~1 1/4          ~2          4 1/4
WHOLE SKY                      332            361            1.253          1.980          4.249

QUASARS OF MAGNITUDE 17

*c*

NUMBER COMPUTED
FROM OBSERVED STATISTICS    ~1/5          ~1/5          ~1/2          ~2/3          ~1
WHOLE SKY                      219            202            560            660            1.026

**NUMBER OF QUASARS** has been estimated by identifying and determining the red shifts of all the quasars in sample fields representing one-thousandth of the whole sky. The sample consisted of 20 quasars with an optical, or apparent, magnitude of about 18. It was clear from their red shifts, however, that some are much farther away than others and therefore are intrinsically brighter, as depicted in *a*. The red-shift intervals have been chosen so that the quasars in any given "shell" of the universe are on the average one magnitude (2.5 times) brighter in absolute luminosity than those in the next shell inward. Thus the four quasars in the sample box representing the most remote shell (red shift: 1.58 to 2.51) are each 100 times more luminous than the single quasar in the box whose red shift is between .16 and .25. Now, if quasars are uniformly distributed in space, and if there are 4,000 of maximum luminosity in the most remote shell, one would expect to find a proportional number in the next shell inward, whose volume is only two-thirds that of the outer shell. Two-thirds times 4,000 is 2,667. Thus diagram *b* shows that in the red-shift interval between 1 and 1.58 one would expect to find 2,667 quasars of maximum luminosity in the whole sky (or, proportionately, 2⅔ quasars in the small area actually sampled). In photographs these 2,667 should appear one magnitude brighter (magnitude 17) than the 4,000 of the same intrinsic brightness that are farther away. Using the same assumptions, one can estimate the number of quasars in still nearer shells.

486        729

1.58                2.51

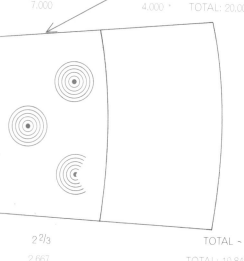

7            4        TOTAL: 20

7,000        4,000        TOTAL: 20,000

2 2/3                TOTAL: ~11

2.667                TOTAL: 10,842

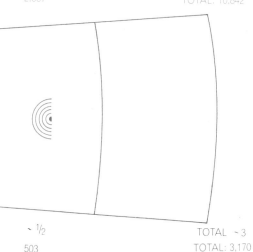

~ 1/2                TOTAL: ~3

503                TOTAL: 3,170

The total number is 10,842, distributed as shown in *b*. One concludes, therefore, that if quasars are uniformly distributed in space, one should observe about twice as many 18th-magnitude quasars as 17th-magnitude quasars. In actuality, however, surveys show that the number of quasars goes up by a factor of about six per magnitude. To satisfy this observation there can be only about 3,000 quasars of the 17th magnitude in the whole sky. An appropriate red-shift distribution for that approximate number is shown in *c*.

light, ruled out the possibility that 3C 273 and 3C 48 were stars in our galaxy. It was proposed that the red shifts are cosmological, which implies that the two objects have to be billions of light-years away and therefore extremely luminous to look as bright as they do in our night sky. They soon became known as quasars. Within the next few years quasars with even larger red shifts were discovered, including some with red shifts of more than 2, or more than six times the largest red shift ever observed for an ordinary galaxy. On the cosmological hypothesis, a red shift of 2 suggests that the light from the object has been traveling for about 80 percent of the age of the universe.

### The Quiet Quasars

Several hundred radio sources have now been identified with starlike objects. Most of the identifications are made on the basis of positions provided by two or more radio telescopes spaced from several hundred meters to several thousand kilometers apart, used as an interferometer. The technique yields a precise measure of the difference in the time required for radio waves from the source to reach each telescope of the interferometer. One can then locate the source with an accuracy of between one second and 15 seconds of arc. Once the search has been narrowed to an optical candidate the final test is to see if its spectrum shows a red shift. More than half of the objects identified on the basis of their radio position usually turn out to be quasars. The spectroscopic test is unambiguous because the maximum red shift ever recorded for a star is .002; the smallest red shift for a quasar identified on the basis of its radio emission is .158 (for 3C 273).

It was noticed early that quasars usually emit rather strongly in the ultraviolet part of the spectrum. In 1964, when radio positions were known with considerably less accuracy than they are today, Ryle and Sandage conceived the idea of using ultraviolet strength as a clue in searching for optical counterparts of radio sources. They used a technique in which a single photographic plate of a star field was exposed to blue light and then was shifted slightly and exposed to ultraviolet radiation. By visual examination it was possible to readily distinguish strong ultraviolet emitters from normal stars.

In 1965 Sandage noted that objects with excess ultraviolet emission were much more plentiful than known radio sources in typical star fields. He soon

discovered that some of these "blue stellar objects" exhibit red shifts that qualify them as quasars even though no radio emission has been detected from them. Most of the other strong ultraviolet emitters turn out to be white-dwarf stars in our own galaxy. Thus only a small fraction of quasars are strong radio emitters. The rest are radio quiet, or virtually so. It may be that a typical quasar is a strong radio emitter for only a small part of its life-span. Alternatively, it may be that relatively few quasars are born to be strong radio sources.

Two years ago Sandage and Willem J. Luyten published photometric analyses of 301 blue objects in seven survey fields. They counted quasar candidates tentatively selected from these blue objects and estimated that in one square degree of the sky (roughly equal to five times the area of the moon) there is, on the average, .4 quasar brighter than magnitude 18.1. They also estimated that there are five quasars per square degree brighter than magnitude 19.4, and that tentatively there are as many as 100 brighter than magnitude 21.4. Over the entire sky they estimated there may be 10 million quasars of the 22nd magnitude or brighter.

The greater the magnitude, of course, the dimmer the object; every increase of five magnitudes (say from the 18th magnitude to the 23rd) corresponds to a decrease by a factor of 100 in brightness. The number of quasars increases steeply with magnitude, by a factor of about six per magnitude. This steep increase is incompatible with a uniform distribution of quasars in space, as we shall see.

### Counting Verified Quasars

The objects isolated by Sandage and Luyten are defined as "faint blue objects [with] an ultraviolet excess." For a detailed statistical study one has to obtain the spectrum of each "faint blue" candidate individually to establish whether or not it is really a quasar. One of the authors (Schmidt) began this task about four years ago, working with several of the star fields examined by Sandage and Luyten. The ultimate goal of the study is to establish how quasars are distributed by red shift (distance) and luminosity.

Of 55 faint blue objects investigated in two of the Sandage-Luyten fields, 32 turned out to have negligible red shifts and therefore could be rejected as being dwarf stars within our own galaxy. The 23 remaining objects exhibited spectra characteristic of quasars, and all but one of the spectra contained the minimum of two lines needed for establishing a red shift. A single line could represent almost any emitting atom red-shifted by any arbitrary amount. When a spectrum contains two lines, however, it is almost always possible to assign a unique red-shift value that identifies a reasonable emission wavelength for each line [see illustration on page 45]. Unfortunately the spectra of some quasars show only a single clear line, thereby frustrating efforts to establish their red shift. Although the red shifts assigned to several of the objects are still tentative, the overall distribution must be essentially correct. The red shifts range from .18 to 2.21. None of the 23 quasars appears in any of the catalogues of strong radio sources.

At this stage it will be most useful in our discussion to concentrate on the quasars of 18th magnitude in the sample. There are 20 such quasars. Since the 20 objects exhibit a variety of red shifts, however, we know they must lie at vastly different distances and therefore must differ greatly in *absolute* luminosity even though they look equally bright to an observer.

To express these differences in absolute luminosity one can classify the objects by red shift in such a way that each red-shift category represents a step of one magnitude in absolute luminosity. The relation between the red shift and the magnitude of a standard source depends on the properties of the universe. In the cosmological model followed in this study a quasar of 18th magnitude whose red shift falls in the range between .25 and .4 is intrinsically brighter by one magnitude than a nearer object whose red shift lies between .16 and .25. Six red-shift categories, each corresponding to a step of one magnitude in absolute luminosity, are enough to cover the range of red shifts actually exhibited by the 20 objects. The brightest members of the group are five magnitudes, or 100 times, brighter than the least luminous.

When the 20 quasars were grouped by red shift in this way, their distribution was found to be similar to the red-shift distribution of radio quasars of the same optical magnitude. Taking this into account and rounding things off somewhat, the following distribution for the red shifts of 18th-magnitude quasars was adopted for the subsequent analysis:

| | | |
|---|---|---|
| Red shift | 1.58–2.51 | 20 percent |
| Red shift | 1.00–1.58 | 35 percent |
| Red shift | .63–1.00 | 20 percent |
| Red shift | .40– .63 | 15 percent |
| Red shift | .25– .40 | 5 percent |
| Red shift | .16– .25 | 5 percent |

The Sandage-Luyten survey had shown that in the entire sky there are, in round numbers, 20,000 quasars of apparent magnitude 18—just 1,000 times as many as in the new detailed sample.

### APPROXIMATE VISUAL MAGNITUDE

| REDSHIFT Z | SHELL VOLUME OF UNIVERSE ($10^{27}$ CUBIC LIGHT YEARS) | SHELL VOLUME OF UNIVERSE $X(1 + Z)^6$ |
|---|---|---|
| 1.58 —2.51 | 729 | 593,000 |
| 1.00 —1.58 | 486 | 74,600 |
| .63 —1.00 | 295 | 10,900 |
| .40 — .63 | 146 | 1,800 |
| .25 — .40 | 61 | 336 |
| .16 — .25 | 22 | 68 |
| .10 — .16 | 7.3 | 15 |
| .06 — .10 | 2.1 | 3.4 |
| .04 — .06 | .59 | .8 |
| .025— .04 | .16 | .19 |
| .016— .025 | .04 | .05 |

**TOTAL QUASAR POPULATION** of universe is estimated to be on the order of 14 million, of which more than 99.7 percent are evidently fainter than the 18th magnitude and have red shifts greater than .4. From the 13th to 18th visual magnitude the number of quasars increases by a factor of five or six

If the 20,000 are distributed according to the percentages listed above, one finds that the number in each red-shift category, starting with the highest, is as follows: 4,000, 7,000, 4,000, 3,000, 1,000 and 1,000. It is clear that in a random sample of 18th-magnitude quasars more than half are extremely distant (red shift greater than 1) and therefore belong to the most luminous members of their class. A red shift of 1 corresponds to looking back two-thirds of the time that has elapsed since the universe began its expansion.

Proceeding to the next stage of the analysis, one would like to estimate the number of quasars whose apparent magnitude is either brighter or fainter than 18 and how they are distributed according to red shift. To do this one must know the volumes of the successive shells of the universe in which we have placed our 18th-magnitude quasars. These volumes depend on the cosmological model followed. Our unit of volume, $10^{27}$ cubic light-years, or a cube of which each side is a billion light-years, is co-moving, which means that no matter what the red shift, the unit of volume expands with the universe into our "local" unit of $10^{27}$ cubic light-years.

The 4,000 brightest and most distant quasars (red shift 1.58 to 2.51) occupy a shell with a volume of $729 \times 10^{27}$ light-years. The problem now is to use this in-

### NUMBER OF QUASARS IN WHOLE SKY

| REDSHIFT Z | SHELL VOLUME OF UNIVERSE ($10^{27}$ CUBIC LIGHT YEARS) | MAGNITUDE 17 CORRECTED DISTRIBUTION ~$(1 + Z)^6$ | MAGNITUDE 17 IF UNIFORMLY DISTRIBUTED | MAGNITUDE 18 DERIVED FROM OBSERVATION |
|---|---|---|---|---|
| 1.58—2.51 | 729 | | | 4,000 |
| 1.00—1.58 | 486 | 503 | 2,667 | 7,000 |
| .63—1.00 | 295 | 1,026 | 4,249 | 4,000 |
| .40— .63 | 146 | 660 | 1,980 | 3,000 |
| .25— .40 | 61 | 560 | 1,253 | 1,000 |
| .16— .25 | 22 | 202 | 361 | 1,000 |
| .10— .16 | 7.3 | 219 | 332 | |
| | | 3,170 | 10,842 | 20,000 |

**DISTRIBUTION OF QUASARS** according to red shift is shown for 20,000 quasars of 18th optical magnitude (*column at far right*), based on a representative sample of 20 quasars. The adjacent columns present two different estimates of the total number of quasars of the 17th magnitude. The method of making the estimates is explained in the illustration on the preceding two pages, where the same numbers appear in the diagrams labeled *b* and *c*. Observation shows that the number of quasars goes up by a factor of about six per magnitude rather than the factor of two expected if quasars were uniformly distributed throughout space. One can obtain the observed distribution by multiplying the shell volume of the universe by $(1 + z)^6$, where $z$ is the red shift and the exponent 6 is an experimentally determined value that yields the desired increment per magnitude. The table at top of these two pages shows the computed number and red shift of all quasars from magnitude 13 through 23.

## APPROXIMATE VISUAL MAGNITUDE

| 13 | 14 | 15 | 16 | 17 | 18 | 19 | 20 | 21 | 22 | 23 |
|---|---|---|---|---|---|---|---|---|---|---|
| - | — | — | — | — | 4,000 | 56,000 | 217,000 | 1,000,000 | 2,000,000 | 9,000,000 |
| - | — | — | — | 503 | 7,000 | 27,000 | 124,000 | 200,000 | 1,000,000 | ——— |
| - | — | — | 74 | 1,026 | 4,000 | 18,000 | 32,000 | 200,000 | ——— | ——— |
| - | — | 12 | 169 | 660 | 3,000 | 5,000 | 27,000 | ——— | ——— | ——— |
| - | 2 | 32 | 123 | 560 | 1,000 | 5,000 | ——— | ——— | ——— | ——— |
| 0 | 6 | 25 | 113 | 202 | 1,000 | ——— | ——— | ——— | ——— | ——— |
| 1 | 6 | 25 | 44 | 219 | ——— | ——— | ——— | ——— | ——— | ——— |
| 1 | 6 | 10 | 50 | ——— | ——— | | | | | |
| 1 | 2 | 12 | ——— | | ——— | | | | | |
| 1 | 3 | — | | | | | | | | |
| 1 | — | | | | | | | | | |
| 5 | 25 | 116 | 573 | 3.170 | 20.000 | 111.000 | 400.000 | 1.400.000 | 3.000.000 | 9.000.000 |

for each decline of one magnitude in brightness. Beyond the 18th magnitude, however, the increase is slower because the table contains no entries for quasars with red shifts greater than 2.51. In fact, only two quasars with larger red shifts are known, which suggests that there is a genuine paucity of such objects. Any quasar with a red shift of 2.5 is so distant that its light has been traveling through space for more than 85 percent of the age of the universe. The light from the more than 13.5 million quasars with a red shift greater than 1 has been en route for at least 6.8 billion years, assuming that the universe is on the order of 10 billion years old. Because the lifetime of a quasar is probably well under a billion years, the overwhelming majority of all the quasars that ever existed must have evolved by now into less luminous objects, perhaps ordinary galaxies. One can estimate that only about 35,000 quasars exist today.

formation to compute how many quasars of the same absolute luminosity would appear in the shell immediately within the outermost one, whose red shift corresponds to between 1 and 1.58. That shell, according to the cosmological model selected, has a volume of $486 \times 10^{27}$ cubic light-years, or two-thirds of the volume of the outer shell. Now we introduce a supposition. If quasars were uniformly distributed in space, the inner shell would contain two-thirds times 4,000, or 2,667, quasars exactly like those in the outer shell. If quasars of that intrinsic luminosity were moved one shell closer to us, their apparent luminosity, as we observe them, would therefore be one magnitude brighter, that is, magnitude 17 instead of magnitude 18 [see illustration on pages 46 and 47]. Remember that the red-shift intervals were chosen specifically so that each step would correspond to a one-magnitude change in brightness.

A similar computation is now performed for the 18th-magnitude quasars in each of the other red-shift categories. In each case one computes the number expected in the shell within the preceding one, assuming as before that quasars are uniformly distributed in space. This calculation yields the following additional numbers: 4,249, 1,980, 1,253, 361 and 332. When these are added to the number 2,667 previously computed, one obtains a total of 10,842 quasars of apparent magnitude 17, or roughly half as many quasars as one expects to find of magnitude 18 (assuming uniform distribution).

We recall that the Sandage-Luyten survey shows that the number of quasarlike objects increases not by a factor of two per magnitude (from 10,842 to 20,000 in the exercise just completed) but by a factor of about six. In other words, their statistics would predict only some 3,000 or 4,000 quasars of apparent magnitude 17 rather than 10,842.

What the factor of six tells us, of course, is that there are more faint quasars than one would expect to find if space were uniformly filled with quasars. The only plausible explanation is that the density of quasars must increase with increasing distance, that is, as we look back farther in time. To arrive at a distribution law that satisfies the observational evidence, let us assume that the density is proportional to some power, $n$, of the scale of the universe. The scale, or size, of the universe is inversely proportional to the amount by which light has been "stretched" by the expansion of the universe. Thus if the Lyman-alpha line emitted at 1,216 angstroms is observed at 3,648 angstroms, one can say that the universe has expanded by a factor of three since the radiation left the emitter. Since the red shift, $z$, in this case is 2 (3,648 minus 1,216 divided by 1,216) it is evident that the scale of the universe is given not by $z$ but by $1 + z$. The density law we are seeking is therefore $(1 + z)^n$.

The value of $n$ is simply obtained by trial and error to yield about 3,000 quasars of magnitude 17 [see illustration on opposite page]. Quite by accident the value of $n$ turns out to be 6. It is only a coincidence that $n$ is 6 and that the increase in the number of quasars per magnitude is also six. With this density law it is a simple matter to extend the distribution table downward from magnitude 17 and upward from magnitude 18 [see illustration above].

Along the bottom of the table one can read off the number of quasars expected in the entire sky for each magnitude. For the five magnitudes brighter than magnitude 18 the expected quasar population decreases steadily at each step from 3,170 (17th magnitude) to 573 (16th) to 116 (15th) to 25 (14th) and finally to five (13th). For the five magnitudes fainter than 18 the expected population rises steeply at each step from 111,000 (19th magnitude) to 400,000 (20th) to 1.4 million (21st) to three million (22nd) and finally to nine million (23rd). The total estimated quasar population from magnitude 13 to magnitude 23 inclusive is thus about 14 million.

The table does not list entries for

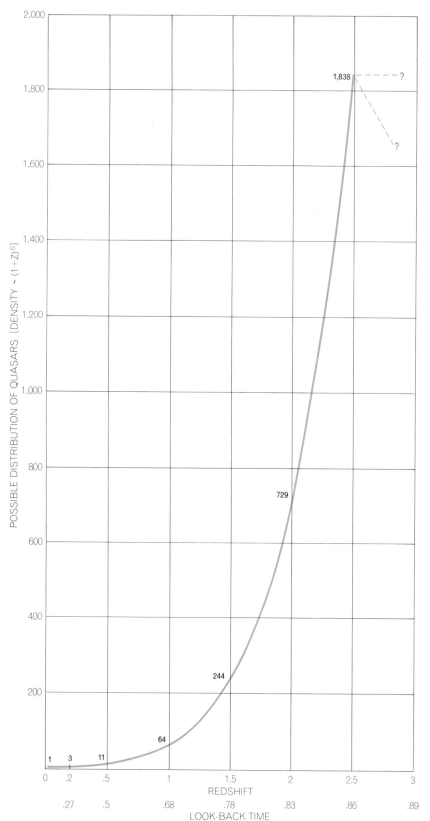

**CHANGE IN QUASAR DENSITY WITH TIME** can be derived from the table on the preceding two pages. The curve shows that the number of quasars rises steeply with increasing red shift, which is equivalent to looking back in time. Thus if one looks back 68 percent of the age of the universe, one would find more than 60 quasars in the volume of space that now contains one quasar. Looking back 83 percent of the age of the universe, one would find more than 700 quasars in the same volume. The maximum density may have existed when the universe had reached only about 14 percent of its present age. The scarcity of quasars with a red shift greater than 2.5 suggests that their density was no greater at earlier epochs.

quasars with red shifts greater than 2.5. Actually two quasars with larger red shifts are known: one, PHL 957, has a red shift of 2.69; the other, 4C 05.34, has a red shift of 2.88. Their magnitudes are respectively 17 and 18. If the density law $(1 + z)^6$ continued to hold, one would expect a great many 19th-magnitude quasars with red shifts larger than 2.5. Their scarcity suggests that the density does not increase beyond 2.5 and that it may actually decrease.

The probable scarcity of quasars with red shifts greater than 2.5 implies that the largest telescopes are able to look back in time to the epoch when quasars made their first appearance in the universe. Depending somewhat on the cosmological model selected, one can say that the light from a quasar with a red shift of 2.5 began its journey through space some 8.6 billion years ago, or some 1.5 billion years after the big bang that hypothetically created the universe as we know it. Within the next few billion years the great majority of quasars were born and began their brief but brilliant career [see illustration at left].

One can estimate that the universe at present contains only some 35,000 quasars. All the rest have presumably evolved into less remarkable objects, perhaps ordinary galaxies; we know of their existence because the signals they emitted billions of years ago are only now reaching our telescopes. The quasars of the lowest intrinsic luminosity (those at the bottom of the table on the preceding two pages) are no brighter than large galaxies. It is therefore uncertain whether all of them are quasars or whether some are compact galaxies of one kind or another. To avoid such confusion one could consider leaving out the quasars listed in the two lowest (least luminous) categories all across the table. The remaining "high luminosity" quasars would then number about 1.5 million for the entire sky, and the number existing at the present time would drop to only 3,500.

Another way to look at the quasar population developed in this analysis is to compare the number of quasars with the number of galaxies in a given volume of space. A volume of $10^{27}$ cubic light-years in our neighborhood contains about 20 quasars, of which two are objects of high luminosity. In very round numbers the same volume of space contains probably between one million and 10 million galaxies.

The study described above involved quasars selected solely on the basis of their optical properties; their radio emission, if any, is negligible. It is therefore important to ask if quasars selected on the basis of their radio luminosity also show an increase in density with increasing distance. The 3C catalogue mentioned above is a comprehensive listing of all radio sources in the northern half of the sky with a certain minimum radio intensity. (The minimum value is nine "flux units" at 178 megahertz, or $9 \times 10^{-26}$ watt per square meter per hertz.) By the late 1960's 44 of the 300-odd extragalactic radio sources in the 3C catalogue had been optically identified as quasars. Of these 44 objects 33 had optical magnitudes of 18.5 or greater, and there was reason to believe that the 33 represented essentially all the 3C quasars down to that limiting magnitude.

## Radio-bright Quasars

The analysis of the distribution of the 33 objects is complicated because both a radio limitation and an optical limitation were involved in their selection. That is to say, to appear in the group of 33 quasars an object had to radiate strongly in two widely separated parts of the spectrum: the radio region and the optical region. The analysis made by one of the authors (Schmidt) went as follows:

From the red shift the distance to each object was computed on the basis of some particular model of the expanding universe. This distance equaled the radius of the volume of space within which the object was actually observed. One can then ask how far the object could be moved outward before one of two things happen: either its apparent magnitude drops below 18.5 or its radio flux falls below nine units. This distance defines the radius of the maximum volume beyond which the object could not lie and still remain a member of its original class.

For each object one can express the ratio of the two volumes, actual volume over maximum volume, as a decimal fraction. A priori, if the 33 objects were uniformly distributed, one would expect the average value of this fraction to be .5. Thus one would expect half of the values to be less than .5 and the other half to lie between .5 and 1. Actually only six of the objects yield values below .5 whereas 27 give higher values. In other words, radio quasars tend to occupy the outer reaches of the volume within which they can be observed. This tells us that their density increases with distance. When the density law is worked out in detail, it is found to lie

between $(1 + z)^5$ and $(1 + z)^6$. That is remarkably similar to the density law obtained for optically selected quasars, which on the average show negligible radio emission. The conclusion is that quasars have a density distribution that is only slightly or not at all dependent on their radio properties. This still leaves unsettled, however, the two possibilities already mentioned: either most quasars pass through a brief evolutionary stage during which they emit strongly at radio wavelengths or else only a small fraction of all quasars are destined to evolve into strong radio emitters.

## Other Quasar Hypotheses

A number of astronomers and theorists originally found it difficult to accept the idea that the red shifts of quasars are cosmological. They did not see how it was possible for an object to emit as much light as 100 galaxies and yet vary in intensity by 10 percent or more in a few days. They proposed, as one alternative, that quasars might be much nearer and smaller objects ejected at high velocity from the center of our own galaxy. This is sometimes called the local-Doppler hypothesis because the red shift is a Doppler shift and the objects are of local origin. Being only a few million light-years away, rather than billions of light-years, their actual energy output would be much less.

This hypothesis has encountered the difficulty that quasars are much more numerous than anyone suspected in the early 1960's. As we have just seen, recent estimates run into the millions, and on the most conservative basis one can hardly assume fewer than a million quasars. It may be estimated that the mass of the typical quasar, on the basis of the local-Doppler hypothesis, would have to be at least 10,000 suns. The ejection of a million objects, each of 10,000 solar masses, from the center of our galaxy would require that the mass of all the stars in the galactic nucleus be completely converted into energy. One must also explain why the only quasars ever observed are those ejected by our own galaxy. If any quasar-like objects had been ejected by any of the scores of galaxies in our immediate neighborhood, some of them should be observed to be heading *toward* us and thus should exhibit a blue shift rather than a red shift. Yet no quasar-like object with a blue shift has ever been detected. The local-Doppler explanation, on the whole, must be regarded as being quite unlikely.

A totally different explanation for the

red shift of quasars seemed attractive at first. According to this hypothesis quasars are objects in which a substantial mass is compressed into an extremely small volume. Light emitted from such an object would have to overcome an immense gravitational potential and would be red-shifted just as it is in quasars. The physical conditions that the hypothesis must account for can be rather precisely calculated. It is possible to compute, therefore, how large an emitting envelope of gas is needed, and what its density and temperature must be, to produce the spectral lines actually observed in quasars.

But if one assumes, to take an extreme case, that the highly condensed mass is comparable to the mass of the sun, its emitting envelope would not exhibit the required luminosity unless it were within 10 kilometers of the observer! The object has to be more distant, of course, and that will require a larger mass. The masses computed are large, and thus tend to create inadmissible side effects. For example, at a distance of 30,000 light-years the mass would have to be $10^{11}$ suns; it would rival the mass of our own galaxy, whose center is at the same distance. If the mass is raised still further to $2 \times 10^{13}$ suns, the minimum distance can be raised to 10 million light-years. In that case, in order not to raise the observed average density of the universe, a million such quasars would have to be distributed out to a distance of at least a billion light-years, at which point they would hardly qualify any longer as local objects.

One other "anticosmological" hypothesis should be mentioned for the sake of completeness: the hypothesis that the cause of the quasar red shift is simply unknown and thus lies outside present-day physics. Since no arguments can be made against such a metaphysical hypothesis it cannot be excluded.

## The Cosmological Hypothesis

An attractive feature of the cosmological hypothesis is that the quasar red shift comes "free," without requiring the introduction of bizarre physical conditions to explain the shift. The quasars exhibit a red shift simply because they are being carried along by the expansion of the universe. The extraordinary luminosity of quasars, together with their short-term variability, originally constituted the strongest objection to the cosmological hypothesis. In the past five years, however, short-term luminosity fluctuations of considerable magnitude have been observed in the nuclei of two rather special kinds of galaxy: N-type galaxies and Seyfert galaxies. These nuclei are starlike and resemble quasars in producing an excess of ultraviolet radiation. Moreover, there is general agreement that their red shifts, even though they are modest in the case of Seyfert galaxies, are cosmological in origin.

Recently it has been found that both quasars and Seyfert galaxies radiate strongly in the infrared region of the spectrum. Indeed, the infrared luminosity of the nearby Seyfert radio galaxy 3C 120 is $10^{46}$ ergs per second, which is equal to the infrared luminosity of many quasars when their luminosity is calculated on the assumption of their being at cosmological distances. In other words, we now have examples of objects whose extraordinary energy output is as difficult to explain as the output of quasars (regarded as cosmological objects) and whose output varies over time scales that are just as brief as the time scales for the variation of quasars. Therefore the cosmological hypothesis cannot be ruled out on the basis of the difficulties encountered in explaining the quasars' rapidly varying high luminosity, because the same difficulties hold for galaxies whose properties and distances are not in question.

Support for the cosmological hypothesis has recently been obtained by James E. Gunn of the Hale Observatories. He found that the image of the quasar PKS 2251 + 11 (red shift .323) is superposed on the image of a small, compact cluster of galaxies. Gunn was able to determine the red shift of the brightest galaxy in the cluster and found a value of .33 ± .01. The coincidence in direction and red shift makes it very likely that the quasar is associated with the cluster of galaxies, thus confirming the cosmological nature of its red shift.

As for the ultimate source of the tremendous energy observed in quasars, there has been no lack of hypotheses, among them stellar collisions, the gravitational collapse of massive stars, supernova explosions, conversion of gravitational energy into particle energy by magnetic fields, matter-antimatter annihilation and the rotational energy of a very compact mass (as proposed for pulsars). There is also no agreement about the radiation mechanism, particularly in the infrared, where much of the output is radiated. Similar problems exist for nuclei of galaxies, notably for those of Seyfert galaxies. The solution of these problems constitutes one of the main challenges to present-day astronomy.

# The Origin of Galaxies

by Martin J. Rees and Joseph Silk
*June 1970*

*The size, shape and other properties of the observed galaxies are traced to slight enhancements in the expanding primordial fireball. Enhancements of certain mass were favored over others*

Perhaps the most startling discovery made in astronomy this century is that the universe is populated by billions of galaxies and that they are systematically receding from one another, like raisins in an expanding pudding. If galaxies had always moved with their present velocities, they would have been crowded on top of one another about 10 billion years ago. This simple calculation has led to the cosmological hypothesis that the world began with the explosion of a primordial atom containing all the matter in the universe. A quite different line of speculation argues that the universe has always looked as it does now, that new matter is continuously being created and that new galaxies are formed to replace those that disappear over the "horizon."

On either hypothesis it is still necessary to account for the formation of galaxies. Why does matter tend to aggregate in bundles of this particular size? Why do galaxies comprise a limited hierarchy of shapes? Why do spiral galaxies rotate like giant pinwheels? Astrophysicists are trying to answer these and similar questions from first principles. The goal is to explain as many aspects of the universe as one can without invoking special conditions at the time of origin. In most of what follows we shall assume a cosmological model in which the universe starts with a "big bang." When we have finished, the reader will see, however, that some form of continuous creation of matter may not be ruled out.

Before the invention of the telescope the unaided human eye could see between 5,000 and 10,000 stars, counting all those visible in different seasons. Even modest telescopes revealed millions of stars and in addition disclosed the existence of many diffuse patches of light, not at all like stars. These extragalactic "nebulas," many of them beautiful spirals, are seen in all directions and in great profusion. As early as the 18th century Sir William Herschel and Immanuel Kant suggested that these nebulas were actually "island universes," huge aggregations of stars lying beyond the limits of the Milky Way.

The validity of this hypothesis was not confirmed until 1924, when the American astronomer Edwin P. Hubble succeeded in measuring the distances to a number of spiral nebulas. Several years earlier Henrietta S. Leavitt had shown that Cepheid variables, named for the prototype Delta Cephei, a variable star discovered in 1784, had light curves that could be correlated with their magnitude. The distances of a number of Cepheids were later determined by independent means, so that it became possible to use more distant Cepheids as "standard candles" to establish a distance-magnitude relation. Hubble looked for Cepheid variables in some of the nearer external galaxies and found them. From their period he was able to deduce their absolute luminosity, and from this he was able to estimate their distance. Hubble soon established that the nearest spiral nebulas (or galaxies) were vast systems of stars situated a million or more light-years outside our own galaxy.

Subsequently Hubble developed a scheme for classifying galaxies according to their morphology, ranging from systems that are amorphous, reddish and elliptical to systems that are highly flattened disks with a complex spiral structure containing many blue stars and lanes of gas and dust [*see illustration on page 55*]. The spiral galaxies themselves vary in appearance. At one extreme are those with large, bright nuclei and inconspicuous, tightly coiled spiral arms. At the other extreme are galaxies in which the nuclei are less dominant and the spiral arms are loosely wound and prominent. The elliptical galaxies also form a sequence, ranging from almost spherical systems to flattened ellipsoids. In addition there are highly irregular systems showing very little structure of any kind.

In all these sequences there is a parallel progression in certain characteristics of the galaxies. In general spirals are rich in gas and dust, contain many blue supergiant stars, are highly flattened and rotate appreciably. Ellipticals, by contrast, seem to possess little gas or dust, usually contain late-type dwarf stars and exhibit scant rotation.

The masses of galaxies are found by several methods. Galaxies are often gravitationally bound together in pairs. If the distance between them and their relative velocities are known, Kepler's law can be used to find their total mass.

CLUSTER OF GALAXIES in the constellation of Hercules demonstrates the inhomogeneity of the distribution of galaxies in the sky. About 350 million light-years away, this cluster contains about 100 members and is some five million light-years across. It was photographed with the 200-inch Hale telescope on Palomar Mountain. Some very rich clusters contain 1,000 members or more and vary from one million to 10 million light-years across. There is some evidence that such clusters are in turn grouped together into superclusters of perhaps 100 members, spread over 100 million light-years. On scales larger than this the universe appears to be uniform. The bright circular spots with the spikes radiating from them are nearby stars; the spikes are produced by reflections within the telescope.

Another method, used mostly for spirals that are viewed edge on or obliquely, is to determine the velocity of rotation by measuring the Doppler shift of spectral lines emitted by ionized gas in various parts of the disk. (The spectral lines of approaching gas will be shifted toward the blue end of the spectrum, those of retreating gas toward the red end.) One can plot a rotation curve showing how the velocity of rotation varies with the distance from the center of the galaxy. The mass can then be estimated from the requirement that the centrifugal and gravitational (centripetal) forces must be in balance. It turns out that the masses of galaxies are typically about $10^{11}$ (100 billion) times the mass of the sun. The range, however, is fairly broad: from about $10^8$ solar masses for some nearby dwarf galaxies to $10^{12}$ solar masses for giant ellipticals in more remote regions of the universe. The diameter of the larger spirals, such as our own galaxy, is about 100,000 light-years.

Galaxies also differ widely in the ratio of mass to luminosity. Taking the mass-to-luminosity ratio of the sun as unity, one finds that for large spirals, such as our own galaxy, the ratio varies from one up to 10. In other words, some spirals emit only a tenth as much light per unit of mass as the sun does. Ellipticals commonly emit even less: only about a fiftieth as much light per unit of mass. (Thus their mass-luminosity ratio is 50.)

The distribution of galaxies in the sky is quite inhomogeneous. There are many small groups of galaxies, and here and there some rich clusters containing up to 1,000 members or more. Such systems vary from one million light-years across to 10 million. Our own galaxy is a member of the "local group," an association of about 20 galaxies, only one of which, the Andromeda galaxy, has a mass comparable to that of ours. The local group is about three million light-years in diameter. The Andromeda galaxy is some two million light-years away; the nearest large cluster of galaxies is in Virgo, about 30 million light-years distant.

Even such clusters do not seem to be randomly distributed in space. Some astronomers have argued that there is evidence that clusters are grouped into superclusters of perhaps 100 members, spread over 100 million light-years. The universe appears to be uniform on scales larger than this.

Establishing the distance of galaxies was only part of Hubble's achievement. Working with the 100-inch telescope on Mount Wilson, he showed from red-shift measurements that the galaxies are in recession. Hubble found, moreover, that the red shift of a galaxy is directly proportional to its distance, as judged by its apparent luminosity. The most distant galaxies known are in a faint cluster in the constellation Boötes; Rudolph Minkowski discovered that the wavelength of light coming from this cluster is stretched by 45 percent. The corresponding velocity of recession is nearly half the speed of light. Light originating from some of the brilliant starlike objects known as quasars is red-shifted more than 200 percent, but astronomers disagree whether or not this red shift is due to the cosmological expansion of the universe.

The light from Minkowski's cluster of galaxies set out toward us about five bil-

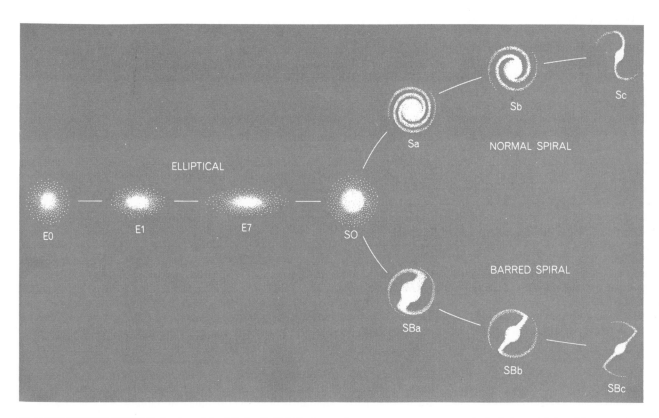

CLASSIFICATION SCHEME developed by Edwin P. Hubble in the early 1930's organizes galaxies according to their shape, ranging from amorphous elliptical systems containing many red stars and little gas and dust (*left*) to highly flattened spiral disks containing many blue stars and lanes of gas and dust (*right*). The elliptical galaxies range from almost spherical systems (designated E0) to highly flattened ellipsoids (E7). The spiral galaxies themselves form two sequences: normal spirals (*top right*) and barred spirals (*bottom right*). At one extreme in both sequences are galaxies with large bright nuclei and inconspicuous, tightly coiled spiral arms (Sa, SBa); at the other extreme are galaxies in which the nuclei are less dominant and the spiral arms are loosely wound and prominent (Sc, SBc). At the branching point of the diagram is a disklike form that resembles the spirals but lacks spiral arms (SO).

lion years ago, and so we can be sure that some galaxies are even older than that. On the other hand, as we have mentioned, all the galaxies must have been tightly packed together no more than 10 billion years ago, based on their present recession velocity. Estimates of the ages of stars suggest that our galaxy, and others like it, are unlikely to be much less than 10 billion years old. Hence we are presented with a remarkable coincidence: most galaxies appear to be about as old as the universe. This implies that galaxies must have formed when conditions in the universe were much different from those now prevailing.

It seems clear, then, that the formation of galaxies cannot be treated apart from cosmological considerations. The dynamics and structure of the universe in the large are beyond the scope of Newtonian physics; it is necessary to use Einstein's general theory of relativity. Because of the complexity of the theory, it is practicable to solve the equations only for cases having special symmetry. Until quite recently the only solutions for an expanding universe were those found in 1922 by the Russian mathematician Alexander A. Friedmann. In his idealized models matter is treated as a strictly uniform and homogeneous medium. The universe expands from a singular state of infinite density, with the rate of expansion decelerating as a consequence of the mutual gravitational attraction of its different parts. The universe may have enough energy to keep expanding indefinitely or the expansion may eventually cease and be followed by a general collapse back to a compressed state. Observations of the actual rate of expansion of the universe at different epochs, as determined by the red shift–luminosity relation of the most distant galaxies, fail to tell us unambiguously whether the expansion will finally stop and be reversed or whether it will continue indefinitely.

The clumping of matter into stars, galaxies and clusters of galaxies in the real universe might seem to make Friedmann's models, based on perfect homogeneity, empty exercises. In actuality the "graininess" we observe in the universe is on such a small scale that Friedmann's solutions remain valid. The reason is that the gravitational influence of local irregularities is swamped by that of more distant matter.

Perhaps the most convincing evidence in support of Friedmann's simple description of the universe was supplied

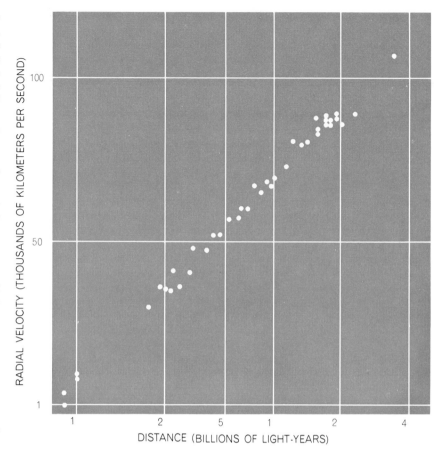

RECESSION VELOCITY OF A GALAXY is obtained by measuring the amount by which the radiation it emits is shifted to the red end of the spectrum. The velocity is directly proportional to the galaxy's distance, as judged by its apparent luminosity. In this diagram, adapted from a recent study by Allan R. Sandage of the Hale Observatories, the ratio of recession velocity to distance is shown for brightest galaxy in each of 41 clusters of galaxies.

in 1965 by the discovery that space is pervaded by a background radiation that peaks at the microwave wavelength of about two millimeters, corresponding to the radiation emitted by a black body at an absolute temperature of three degrees (three degrees Kelvin). This radiation could be the remnant "whisper" from the big bang of creation. The remarkable isotropy, or nondirectionality, of this radiation is impressive evidence for the isotropy of the universe.

The radiation was discovered independently and almost simultaneously at the Bell Telephone Laboratories and at Princeton University [see "The Primeval Fireball," by P. J. E. Peebles and David T. Wilkinson; SCIENTIFIC AMERICAN, June, 1967]. The radiation has the spectrum characteristic of radiation that has attained thermal equilibrium with its surroundings as a result of repeated absorption and reemission, and it is generally interpreted as being a relic of a time when the entire universe was hot, dense and opaque. The radiation would have cooled and shifted toward longer wavelengths in the course of the universal expansion but would have retained a thermal spectrum. It thus constitutes remarkably direct evidence for the hot-big-bang model of the universe first examined in detail by George Gamow in 1940.

Assuming the general validity of the Friedmann model for the early stages of the universe, it seems clear that the material destined to condense into galaxies cannot always have been in discrete lumps but may have existed merely as slight enhancements above the mean density. There will be a tendency for the larger irregularities to be amplified simply because, on sufficiently large scales, gravitational forces predominate over pressure forces that tend to oppose collapse. This phenomenon, known as gravitational instability, was recognized by Newton, who, in a letter to Richard Bentley, the Master of Trinity College, wrote:

"It seems to me, that if the matter of our sun and planets, and all the matter of the Universe, were evenly scattered through all the heavens, and every particle had an innate gravity towards all the rest, and the whole space throughout which this matter was scattered, was finite, the matter on the outside of this space would by its gravity tend towards all the matter on the inside, and by consequence fall down into the middle of the whole space, and there compose one great spherical mass. But if the matter were evenly disposed throughout an infinite space, it could never convene into one mass; but some of it would convene into one mass and some into another, so as to make an infinite number of great masses, scattered great distances from one to another throughout all that infinite space. And thus might the sun and fixed stars be formed, supposing the matter were of a lucid nature."

Newton envisaged a static universe, but the same qualitative picture occurs in an expanding Friedmann universe, as was shown by the Russian physicist Eugene Lifshitz in 1946.

Because of the atomic nature of matter the early universe could never have been completely smooth. It would obviously be gratifying if the inevitable random irregularities in the initial distribution of atoms sufficed ultimately to produce the bound systems of stars we see throughout the universe today. Unfortunately this type of statistical fluctuation fails by many orders of magnitude to account for the observed degree of structure in the universe. Moreover, it remained a puzzle why agglomerations of a certain mass, notably galaxies, should be so plentiful. It appeared necessary to postulate initial fluctuations in a seemingly *ad hoc* manner, and nothing had really been explained; "things are as they are because they were as they were."

Only in the past two or three years has it been realized that the background radiation acts as a gigantic homogenizer on certain preferred scales. To understand just how this works we must look more closely at Gamow's model of the universe. In the early stages, when the universe consisted of a primordial fireball, no structures such as galaxies or stars could have existed in anything like their present form. All space would have been filled with radiation (photons) and hot gas, consisting of the nuclei of hydrogen and helium and the accompanying electrons. The photons would be repeatedly scattered from the electrons. For at least the first 100,-000 years of its history (beginning roughly 10 seconds after its emergence

SPHERICAL GALAXY, classified as type E0 in Hubble's scheme, is a member of the Virgo cluster of galaxies. A representative of the most massive type of galaxy, this system, designated M 87, contains about 30 times as many stars as a spiral system such as our own does.

ELLIPTICAL GALAXY in the constellation of Cassiopeia is a member of the local group of galaxies. Designated NGC 147, it is an E4, or intermediate, type of elliptical galaxy. Because of its comparative proximity to our system it can be resolved into individual stars.

from the initial singularity) the universe can be pictured as a composite gas in which some of the "atoms" are particles and the rest are photons. For the universe as a whole there are now at least 10 million times more photons than particles. From thermodynamic considerations one can conclude that photons must also have greatly outnumbered particles in the fireball. For a gas in equilibrium each species of particle contributes to the total pressure in proportion to its number. This still holds (very nearly) for photons, so that the radiation would make an overwhelmingly dominant contribution to the pressure. (During the first 10 seconds, when the temperature exceeds a few billion degrees, the situation is less simple because pairs of photons can interact to form an electron and a positron.)

As the expansion proceeds and the density decreases, the photons lose energy, the temperature drops and the particles move less rapidly. A key stage is reached after about $10^5$ years, when the fireball has cooled to 3,000 degrees. The electrons are then moving so slowly that virtually all are captured by nuclei and retained in bound orbits. In this condition they can no longer scatter photons and the universe becomes transparent. Inasmuch as the background temperature today is only about three degrees absolute, one can conclude that the universe has expanded by a red-shift factor of 1,000 since the scattering stopped.

(Wavelength is inversely proportional to temperature.)

The microwave background photons have probably propagated freely since the universe became transparent and therefore they should carry information about a "surface of last scattering" at a red shift of more than 1,000. Compare this with the red-shift factor of about one-half for the most distant galaxy known! Because these photons have been traveling unimpeded since long before galaxies existed, they should provide us with remarkably direct evidence of physical conditions in the early universe.

Let us return now to the epoch of the primordial fireball and ask: How were inhomogeneities in the fireball affected by the presence of the intense radiation field? Radiation would inhibit the process of gravitational collapse. Under radiation pressure nonuniformities in the fireball would take the form of oscillations, pressure waves or turbulence. These disturbances, in turn, will be dissipated by viscosity and the development of shock waves. Some wavelengths will be attenuated more severely than others, so that inhomogeneities of favored size will be preserved whereas those less favored will tend to be destroyed. The aim of recent work has been to determine what scales of perturbation are most likely to survive the various damping processes until the scat-

tering of photons comes to an end. Any perturbation whose survival and growth is specially favored should eventually dominate, almost irrespective of how nonuniformities were initially distributed in the primordial fireball. An encouraging result that has already emerged from these studies is that $10^{12}$ solar masses, roughly the mass of a large galaxy, is one such preferred scale [*see illustration on page 61*].

After the electrons in the initial plasma have been bound into atoms, radiation no longer affects the distribution of mass. At this point the surviving perturbations are free to amplify gravitationally. (It should be noted, however, that on small scales—less than $10^6$ solar masses—the kinetic energy of atoms exerts a pressure of its own that inhibits gravitational collapse.) The first generation of bound systems will therefore condense from whatever scale of fluctuations had the largest amplitude at the time of decoupling, that is, when the fireball ceased to be a plasma of electrons and other particles.

At what stage did protogalaxies stop expanding and separate out from the rest of the universe? We might guess that this happened when the mean density was comparable to the present density in the outlying parts of galaxies. In 1962 Olin J. Eggen, Donald Lynden-Bell and Allan R. Sandage of the Hale Observatories investigated the likely early history of our own galaxy by studying

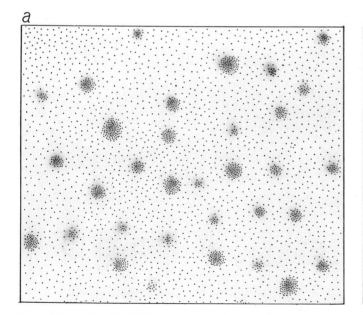

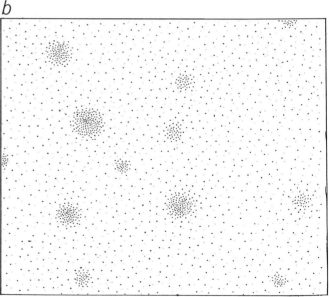

**FORMATION OF GALAXIES** is represented in this sequence of drawings in terms of the "big bang" cosmological model first examined in detail by George Gamow in 1940. For roughly the first 100,000 years after the explosion of the primordial atom the temperature of the expanding fireball was so high that all matter (*black stippling*) was ionized, that is, dissociated into electrically charged particles (*a*). In this situation photons of radiation could not travel very far without being scattered by the free electrons; as a result the universe during this period was effectively opaque (*light shade of color*). Nevertheless, slight random enhancements in the density of matter above the mean density presumably took place, usually accompanied by corresponding enhancements in the

very old stars in the galactic halo. These stars probably formed while the galaxy was collapsing to its present disklike shape (and before the birth of the stars in the Milky Way), and their orbits indicate that our galaxy attained a maximum radius of about 100,000 light-years. One can then tentatively estimate that galaxies such as our own formed when the universe was 1,000 times denser than it is now, about half a billion years after the expansion began.

Extrapolating backward in time, we find that the protogalaxies would have taken the form of nonuniformities roughly 1 percent denser than the average density of the universe at the decoupling epoch. It is an attractive possibility that these are the dominant surviving irregularities, all smaller scales having been smoothed out during the fireball phase. There are, however, some types of fluctuation that are not eradicated in the fireball, so that smaller gas clouds may have formed first and later collided and agglomerated into galaxies. Robert H. Dicke and P. J. E. Peebles of Princeton have suggested that globular clusters—compact groups of about $10^5$ or $10^6$ stars that orbit around galaxies—may represent that small fraction of clouds which managed to avoid collisions, fragmented into stars and survived. Clusters of galaxies would have evolved from initial irregularities of smaller amplitude but larger scale than those destined to form single galaxies.

The only contribution of cosmologists to date toward explaining galaxy formation has been to calculate what scales of perturbation are most likely to survive or amplify in the fireball, thereby reducing the need to build these preferred scales into the initial conditions. This removes one element of arbitrariness in the initial conditions prescribed for the universe. There still remains, however, the task of explaining both the origin of the nonuniformity of the universe on all scales except the very largest, and the apparent uniformity encountered on the largest scales.

In fact, the Friedmann models may not provide an adequate description of the fireball when large inhomogeneities are present. It would be conceptually attractive if there were processes that could transform an initially chaotic universe into one that displayed the large-scale uniformity of a Friedmann model. An encouraging step toward this goal has been taken by Charles W. Misner of the University of Maryland, who has considered a "mix master" universe, which expands anisotropically in such a way that all parts of the universe are causally related very early in its history. At the outset matter would be so densely packed that even neutrinos would interact with other particles at a significant rate. Acting like a blender, the neutrinos would destroy the original anisotropy of the fireball by the time it had cooled to about 20 billion degrees. Thereafter the

expansion would mimic a homogeneous Friedmann model.

Several types of observation may help to test this general picture of galaxy formation. The fluctuations that develop into galaxies and clusters would give rise to random motions on the surface of last scattering. As a result the microwave background photons would not all have been red-shifted by exactly the same amount; in some directions they might have been scattered off material with a random velocity toward us, whereas in other directions the last-scattering surface may have been receding from us. As a consequence the microwave temperature would be slightly nonuniform over the sky. Edward R. Conklin and R. N. Bracewell of Stanford University, Arno A. Penzias, Johann B. Schraml and Robert W. Wilson at the Kitt Peak National Observatory and Yuri N. Parijsky of the Pulkovo Observatory can detect temperature fluctuations as small as a tenth of a percent on angular scales of a few minutes of arc, but so far they have found no positive effect. This technique, however, has the potentiality of detecting embryonic galaxies or clusters of galaxies when they were merely small enhancements above the mean gas density.

There are reasons to expect galaxies that have just condensed to be brighter than typical galaxies at the present epoch. The energy released by the collapse of the protogalaxy would probably have been radiated away by hot gas be-

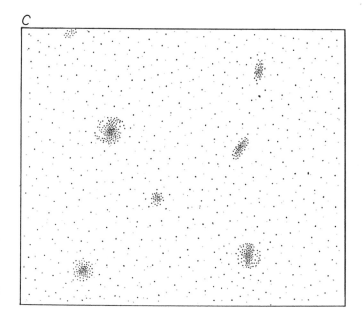

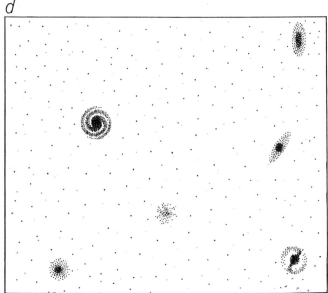

radiation density (adiabatic fluctuations). In such regions (*dark shade of color*) the radiation tended to damp fluctuations that would lead to further enhancements of matter if they were below a certain critical size (about $10^{11}$ solar masses). After about 100,000 years, when expanding fireball had cooled to about 3,000 degrees Kelvin, the negatively charged electrons were moving slow-

ly enough to be captured by protons and retained in bound orbits, forming hydrogen atoms. In this condition electrons are much less effective in scattering photons and universe thus became transparent (*b*). Expansion and cooling of fireball continued and matter was progressively concentrated by gravitational forces, first into protogalaxies (*c*) and eventually into galactic types seen today (*d*).

**SPIRAL GALAXY M 101** in Ursa Major is representative of the Sc type, which is characterized by a relatively inconspicuous nucleus and prominent, loosely wound spiral arms. Our own Milky Way galaxy is either of this type or of the slightly less open Sb type.

**BARRED SPIRAL GALAXY NGC 1300** in Eridanus is classified SBb, which means that it is an intermediate type on the barred-spiral branch of the Hubble sequence. All photographs shown on this page and page 57 were made with the 200-inch Hale telescope.

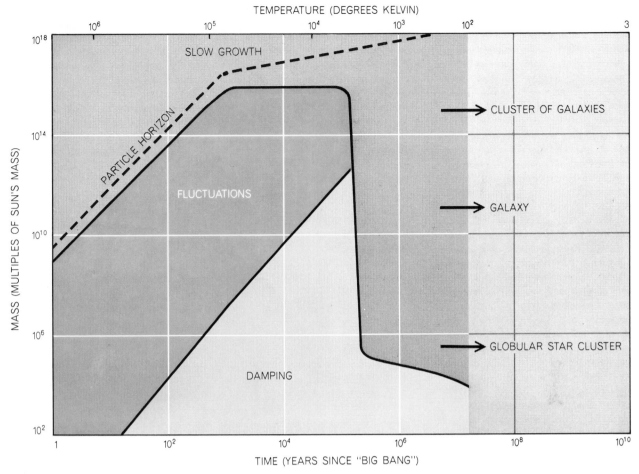

ANOTHER REPRESENTATION of the formation of galaxies is a graph relating the mass of a system to the age and ambient temperature of the universe. Any density enhancement that reaches a minimum value of some $10^{15}$ solar masses when the universe is about 1,000 years old (the epoch at which the density of matter first equals the mass density of the fireball radiation) has enough gravitational force to overwhelm the effects of radiation pressure. Such an enhancement thereupon enters on a lifetime of slow growth, culminating in a large cluster of galaxies. In an intermediate range (between $10^{11}$ and $10^{15}$ solar masses at decou- pling) fluctuations in density persist until the decoupling stage is reached and radiation pressure ceases to interact effectively with matter; surviving density enhancements in this range become in- dividual galaxies. Below a certain threshold ($10^{11}$ solar masses at decoupling) radiation pressure damps out most density enhance- ments. Within this range, however, some density enhancements not accompanied by increases in radiation pressure (isothermal fluctu- ations) may survive to form globular star clusters, ranging from $10^5$ to $10^6$ solar masses. "Particle horizon" is boundary of the observ- able universe, where objects would be receding at speed of light.

fore most of the stars formed. Moreover, the first generation of stars would tend to be heavier and more luminous in rela- tion to their mass than the stellar popu- lations in present-day galaxies. Although most of this energy would be radiated in the ultraviolet, it would be received in the near infrared, owing to the red shift. Robert Bruce Partridge and Peebles at Princeton have suggested that it might be feasible to detect such young galaxies even though these may now have red shifts of about 10.

We are plainly still far from under- standing even the broad outlines of the processes whereby the observed aggregations of matter in the universe came into being. We are even further from understanding the detailed mor- phology of the bewildering variety of different types of galaxies. For example,

we have not yet discussed the possible origins of the angular momentum or magnetic fields of galaxies. Peebles has argued that the rotation of galaxies may be induced by tidal interactions soon after formation. Other authors, notably Leonid Ozernoi of the P. N. Lebedev Physical Institute in Moscow, have con- sidered galactic rotation to be of pri- mordial origin. One remarkable feature of the primordial fireball is that it can store rotation in the form of "photon whirls"; subsequently this stored rotation could be transferred to matter whirls.

Galactic magnetic fields may be pro- duced after the formation of the galaxy by a mechanism of the dynamo type. Alternatively, magnetic fields may very well be of primordial origin. Edward R. Harrison of the University of Massachu- setts has pointed out that the shear be- tween the photon gas and the matter gas

in the fireball could have generated a small magnetic field; if primordial pho- ton whirls are assumed to be present, this mechanism leads to the production of a "seed" magnetic field many orders of magnitude below the value of the magnetic field observed in the spiral arms of our own galaxy. Harrison argues that rapid rotation of the protogalaxy may have subsequently produced suffi- cient winding of the primordial magnetic field to enhance it by the dynamo mecha- nism to the field currently observed. Primordial fields alone, he feels, would be insufficient to account for the ob- served galactic fields. The amount of ro- tation and the strength of the magnetic field in the protogalaxy probably help to determine whether it will evolve into an elliptical galaxy or into a spiral.

Galaxies are observed to possess ran- dom velocities with respect to the cosmic expansion. It is a curious coincidence that the rotational velocity of galaxies is of just the same order of magnitude—hundreds of kilometers per second—as these random motions. Perhaps this is simply a consequence of the primordial turbulence, which may have been the source of all structure in the universe.

Present data on the sizes of clusters of galaxies, and on possible "superclusters," are too sparse to enable us to assess the validity of theories that predict the mass spectrum of condensations. Moreover, our knowledge of the masses of galaxies is bedeviled by selection effects. Large and bright galaxies can be seen out to great distances, but small and intrinsically faint ones would only be noticed if they were comparatively close to us. Such objects may therefore occur much more frequently than is believed. A more drastic possibility is that most of the material in the universe may be in some nonluminous form. Evidence for the existence of such material comes from studies of the stability of clusters of galaxies.

This basic problem was first discussed in 1933 by Fritz Zwicky of the California Institute of Technology. For example, if one estimates the mass required to make the Virgo cluster a gravitationally bound system, one finds that the total observed mass in the member galaxies falls short by a factor of 50 or more. One possible way around this paradox is to assume that the Virgo system may be exploding, as the Soviet astrophysicist V. A. Ambartsumian has suggested. Perhaps even more puzzling is the apparent deficiency in mass of the Coma cluster. This system is so spherically symmetric and centrally condensed that astrono-

mers believe it must be a stable system. Yet the observed mass, predominantly in elliptical galaxies, falls short of the mass required for stability by perhaps a factor of five, even if one assumes that the mass-luminosity ratio for ellipticals is around 50.

Similar results have been found for other clusters. Some astronomers have attempted to explain this problem by arguing that nonluminous matter is present in sufficient quantity to stabilize these systems. This material probably cannot all be in gaseous form; neutral hydrogen or ionized hydrogen, whether uniformly distributed or in clouds, ought to be observable either by radio or by X-ray observations.

Alternatively, the "missing mass" may be in the form of "dead," or burned-out, galaxies. An even more intriguing possibility is that concealed within the clusters are many objects that have undergone catastrophic gravitational collapse, as predicted by the general theory of relativity. The gravitational field around such objects would be so strong that no radiation could escape from them; only their gravitational influence could be detected by a distant observer.

Other arguments that indicate the apparent youthfulness of some galaxies stem from observations of clusters of galaxies. To be stable, one such chain of galaxies would require a mass-luminosity ratio of more than 5,000, or 100 times as much mass as the cluster seems to possess. One seems forced to the conclusion that here are newly formed galaxies, born within the past 100 million years. Zwicky has discovered an entire class of compact galaxies whose surface brightness resembles that found only in the

nuclei of ordinary galaxies. Even more baffling is the discovery that some quasars emit as much radiation as 1,000 galaxies, the energy apparently coming from a colossal explosive event in a region less than 1 percent the size of the solar system. Seyfert galaxies display the same energetic phenomenon on a somewhat reduced scale.

Ambartsumian has long maintained that galactic nuclei are sources of matter and that indeed the galaxies themselves emerge out of dense primordial nuclei. In recent years Halton C. Arp of the Hale Observatories and Erik B. Holmberg of the University of Uppsala have found evidence that small galaxies may even have been ejected from larger galaxies. These phenomena certainly suggest that violent events, involving perhaps the birth of galaxies, are continually taking place in the nuclei of existing galaxies. One is reminded of Sir James Jeans's prescient conjecture, written in 1929, that "the centers of the nebulae are of the nature of 'singular points,' at which matter is poured into our Universe from some other, and entirely extraneous, spatial dimension, so that, to a denizen of our Universe, they appear as points at which matter is being continually created."

Further progress in this field must await fuller information on the distribution, masses and velocities of galaxies. Moreover, satellite observations in infrared, ultraviolet and X-ray wavelengths may soon reveal completely new and unsuspected types of objects, and should in any case give us confidence that we have a fairly complete inventory of the contents of the universe. We shall then be better able to relate theoretical abstractions to the universe in which we dwell.

CHAIN OF GALAXIES VV 172 was photographed by Halton C. Arp with the 200-inch Hale telescope. Four of the galaxies are 600 million light-years away; the fifth appears to be twice as distant. Conceivably it has been ejected from the cluster at high velocity.

EXPLODING GALAXY NGC 1275 was recently photographed in red light by C. Roger Lynds with the 84-inch telescope at Kitt Peak. The radiating filaments of gas, reminiscent of the Crab nebula, were not visible in earlier photographs taken in white light.

# The Search for Black Holes

**8**

by Kip S. Thorne
*December 1974*

*Observations at the wavelengths of light, radio waves
and X rays indicate that the X-ray source Cygnus X-1
is probably a black hole in orbit around a massive star*

Of all the conceptions of the human mind from unicorns to gargoyles to the hydrogen bomb perhaps the most fantastic is the black hole: a hole in space with a definite edge over which anything can fall and nothing can escape; a hole with a gravitational field so strong that even light is caught and held in its grip; a hole that curves space and warps time. Like the unicorn and the gargoyle, the black hole seems much more at home in science fiction or in ancient myth than in the real universe. Nevertheless, the laws of modern physics virtually demand that black holes exist. In our galaxy alone there may be millions of them.

The search for black holes has become a major astronomical enterprise over the past decade. It has yielded dozens of candidates scattered over the sky. At first the task of proving conclusively that any one of them is truly a black hole seemed virtually impossible. In the past two years, however, an impressive amount of circumstantial evidence has been accumulated on one of the candidates: a source of strong X-ray emission in the constellation Cygnus designated Cygnus X-1. The evidence makes me and most other astronomers who have studied it about 90 percent certain that in the center of Cygnus X-1 there is indeed a black hole.

Before I describe the evidence that leads to this conclusion, let me lay some groundwork and indulge my theoretical proclivities by describing some of the predicted properties of black holes [see "Black Holes," by Roger Penrose; SCIENTIFIC AMERICAN, May, 1972]. Physicists educate themselves and their students by means of "thought experiments" whose results are predicted by theory. I shall resort to such an experiment to convey the basic reasoning that underlies the concept of the black hole.

Imagine that at some distant time in the future the human species has migrated throughout the galaxy and is inhabiting millions of planets. Having no further need for the earth, men choose to convert it into a monument: They will squeeze it until it becomes a black hole. To do the squeezing they build a set of giant vises, and to store the necessary energy they fabricate a giant battery. They then scoop out a chunk of the earth and convert its mass into pure energy, the amount of energy obtained being given by Einstein's equation $E = mc^2$, in which $E$ is the energy, $m$ is the mass and $c$ is the speed of light. This energy is stored in the battery. The vises are arrayed around the earth on all sides and, powered by the battery, they squeeze the earth down to a quarter of its original size.

To check their progress the project engineers fabricate from a chunk of the earth a tight-fitting spherical jacket strong enough to hold the planet in its compressed state. They slip the jacket over the earth and open the vises. Then they measure the escape velocity of a rocket placed on the earth, that is, the velocity the rocket must attain in order to be able to coast out of the earth's gravitational field. Before the earth was compressed the escape velocity was the same as it is today: 11 kilometers per second. The compression of the earth, however, brings the earth's surface four times closer to its center, thereby quadrupling the kinetic energy that the rocket must have in order to escape. The escape energy is proportional to the square of the escape velocity; therefore the escape velocity after this first compression is doubled to 22 kilometers per second.

Satisfied that some progress has been made, the engineers repeat the process, compressing the earth still further until its original circumference of 40,000 kilometers is only 10 kilometers. I give the measurement of circumference in-

⟶

MODEL FOR A BLACK HOLE IN CYGNUS X-1 is a likely explanation for the observations made in the visible and X-ray regions of the spectrum. Gas is being pulled off the supergiant primary star HDE 226868 (*a*) by the gravitational attraction of the black hole. As the gas falls toward the black hole, the hole moves in its orbit out of the way, causing the gas to miss it. The gas nearest the black hole is whipped around it into a tight circular orbit, forming a thin accretion disk. The second illustration (*b*), at a scale of about 20 times smaller than the first, shows the expected shape of the accretion disk. The gravitational pull of the black hole compresses the disk, making it thin. At the same time thermal pressures in the gas react against the compression and try to thicken the disk. Only in the central bulge (*c*) are the pressures sufficient actually to thicken the disk. The large pressures in the bulge are caused by heat from X rays emitted near the black hole. In the core of the accretion disk (*d*) the thermal pressures are even higher than they are in the bulge; the gravity is so enormously strong, however, that it prevents the disk from thickening. The X rays observed from the earth are generated only in the innermost 200 kilometers of the core (*e*), which has the black hole itself at its center. In the innermost 50 or 100 kilometers the disk becomes translucent, violently turbulent and much hotter than it is elsewhere. The disk terminates near the black hole, where the gravitational field becomes so strong that gas can no longer move in an orbit but is sucked directly in. The termination point and the structure of the inner disk are sensitive to the black hole's speed of rotation (*see illustration, page 68*). Here it is assumed that the black hole is rotating very rapidly; if the rotation is slow, the X-ray-emitting region may be 400 kilometers in radius rather than 200.

64

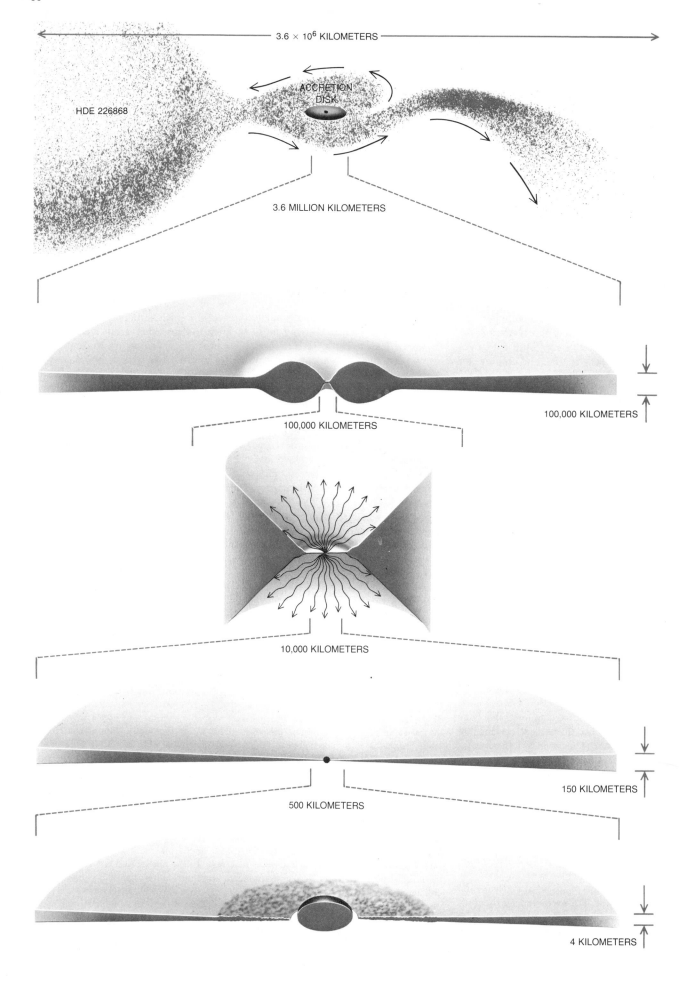

3.6 × 10⁶ KILOMETERS

ACCRETION
DISK

HDE 226868

3.6 MILLION KILOMETERS

100,000 KILOMETERS

100,000 KILOMETERS

10,000 KILOMETERS

500 KILOMETERS

150 KILOMETERS

4 KILOMETERS

stead of diameter because in the presence of strong gravitational fields space is so highly curved that the object's diameter ($d$) is no longer related to its circumference ($C$) by the Euclidean formula $C = \pi d$; moreover, in the case of a black hole the diameter cannot be measured or calculated. This time the rocket needs an escape velocity of 708 kilometers per second in order to coast away from the earth.

After several more compression stages the earth has been reduced to a circumference of 5.58 centimeters. The escape velocity is now 300,000 kilometers per second–the speed of light. One last little squeeze, and the escape velocity exceeds the speed of light. Now light itself cannot escape from the earth's surface, nor can anything else. Communication between the earth and the rest of the universe is permanently ruptured. In this sense the earth is no longer part of the universe. It is gone, leaving behind it a hole in space with a circumference of 5.58 centimeters. Outside the horizon, or edge, of the hole the escape velocity is less than the speed of light, and exceedingly powerful rockets can still get

away. Inside the horizon the escape velocity exceeds the speed of light and nothing can escape. The interior of the hole, like the earth that gave rise to it, is cut off from the rest of the universe.

Let us return to the present and use the thought experiment to aid in understanding what happens in a star. There is a key difference between the earth and a massive star. For the earth to become a black hole external forces must be applied; for a star to become a black hole the necessary forces are provided by the star's own internal gravity. When a star of, say, 10 times the mass of the sun has consumed nuclear fuel through its internal thermonuclear reactions for a period longer than a few tens of millions of years, its fuel supply runs out. With its fires quenched the star can no longer exert the enormous thermal pressures that normally counterbalance the inward pull of its gravity. Gravity wins the tug-of-war, and the star collapses.

Unless the star sheds most of its mass during the collapse, gravity crushes it all the way down to a black hole. If, however, the star can eject enough ma-

terial to reduce its mass to about twice the mass of the sun or less, then it is saved: nonthermal pressures, such as the electron pressures that make it difficult to compress rock, build up and halt the collapse. The star becomes either a white dwarf about the size of the earth or a neutron star with a circumference of some 60 kilometers. (A neutron star is a star where matter is so dense that its electrons have been squeezed onto its protons, converting them into neutrons.) In either case, as with the earth, to convert the object into a black hole one must apply external forces–forces that do not exist in nature [see "Gravitational Collapse," by Kip S. Thorne; SCIENTIFIC AMERICAN, November, 1967].

These predictions, which follow from the standard laws of physics, tell us that there is a critical mass for compact stars (stars with a circumference smaller than the earth's) of about two times the mass of the sun. Below the critical mass a compact star can be a white dwarf or a neutron star. Above the critical mass it can only be a black hole. The magnitude of the critical mass is a key link in the arguments that Cygnus X-1 is a black hole.

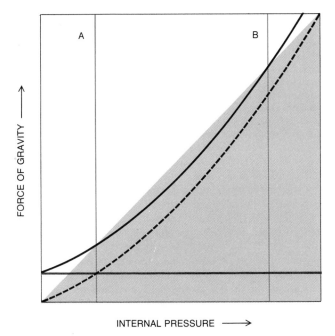

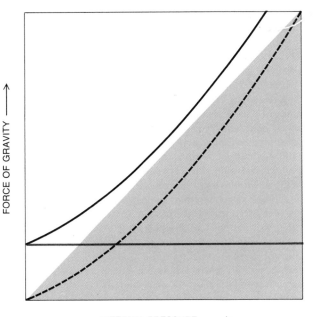

ONLY MASSIVE STARS BECOME BLACK HOLES, as is shown by these diagrams for a star with twice the mass of the sun (*left*) and a star with four times the mass of the sun (*right*). For a star to be stable the total force of gravity (*solid black line*) pulling a star's surface inward must be balanced by the star's internal pressures (*region in light color*) pushing the surface outward. If the force of gravity is stronger, the star collapses; if the internal pressure is stronger, the star explodes. For a compact star (a star smaller than the earth) the internal pressure is large enough to counterbalance gravity if the star's mass is less than three times the mass of the sun. The result is a stable white-dwarf star or a neutron star (a star in which matter is so dense that its electrons have been squeezed onto its protons, converting them into neutrons). If the star is more massive than three times the mass of the sun, however, and has a density greater than the density of the atomic nucleus, the internal pressure actually works against it. According to the general theory of relativity, gravity is produced not only by mass (*gray line*) but also by pressure (*broken black line*). At high pressures the force of gravity generated by the pressure is proportional to the square of the pressure. Thus if the star's internal pressure is very high, it gives rise to gravitational forces that overwhelm the internal pressure and the star collapses. For a compact star with a mass of less than three times the mass of the sun (*left*) there are intermediate pressures that can be counterbalanced by gravity (*vertical lines in color*). For stars more massive (*right*), however, force of gravity always wins and crushes star into a black hole in less than a second.

Therefore one would like to know the critical mass precisely. Precision is not possible, however, because we do not know enough about the properties of matter at the "supernuclear" densities of a white dwarf or a neutron star, that is, at densities above the density of the atomic nucleus: $2 \times 10^{14}$ grams per cubic centimeter. Nevertheless, an upper limit on the critical mass is known: Remo Ruffini of Princeton University and others have shown that it cannot exceed three times the mass of the sun. In other words, no white dwarf or neutron star can have a mass greater than three times the mass of the sun.

From a physical and mathematical standpoint a black hole is a marvelously simple object, far simpler than the earth or a human being. When a physicist is analyzing a black hole, he need not face the complexities of matter, with its molecular, atomic and nuclear structure. The matter that collapsed in the making of the black hole has simply disappeared. It exerts no influence on the hole's surface or exterior. It makes no difference whether the collapsing matter was hydrogen, uranium or the antimatter equivalents of those elements. All the properties of the black hole are determined completely by Einstein's laws for the structure of empty space.

Exactly how simple black holes must be has been discovered by three physicists: Werner Israel of the University of Alberta and Brandon Carter and Stephen Hawking of the University of Cambridge. They have shown that when a black hole first forms, its horizon may have a grotesque shape and may be wildly vibrating. Within a fraction of a second, however, the horizon should settle down into a unique smooth shape. If the hole is not rotating, its shape will be absolutely spherical. Rotation, however, will flatten it at the poles just as rotation slightly flattens the earth. The amount of flattening and the precise shape of the flattened hole are determined completely by its mass and its angular momentum (speed of rotation). The mass and angular momentum not only determine the hole's shape; they also determine all the other properties of the hole. It is as though one could deduce every characteristic of a woman from her weight and hair color.

In calculations the angular momentum is replaced by a more convenient quantity: the rotation parameter. The rotation parameter ($a$) is equal to the speed of light ($c$) multiplied by the angular momentum ($J$), divided by the Newtonian gravitational constant ($G$) times

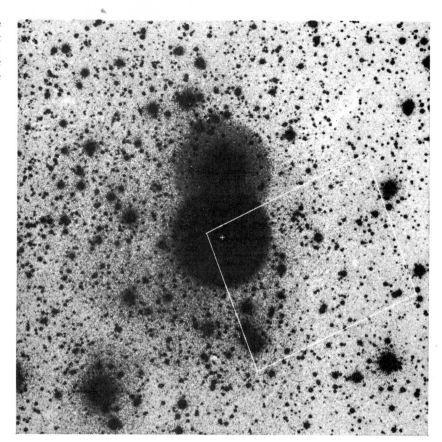

CYGNUS X-1 is believed to be associated with the star HDE 226868, the darkest and largest object in the center of this negative print of a small region of the sky. The photograph was made by Jerome Kristian with the 200-inch reflecting telescope on Palomar Mountain. The photographic plate was exposed so long that exceedingly faint stars (down to the 22nd magnitude) are visible. Superimposed on photograph are location of sources of radio emission (*small cross*) and of X-ray radiation (*white outline*). Position of Cygnus X-1 is not known very well from X-ray observations alone because X-ray telescopes have low resolution. During last week of March and first week of April in 1971, however, Cygnus X-1 underwent a cataclysmic and so far permanent change that caused it to begin emitting radio waves and to double the average energy of its X rays. Because the positions of radio sources can be measured accurately, the change in Cygnus X-1 assisted astronomers in identifying its location.

the square of the hole's mass ($m^2$): ($a = c J/G m^2$). The rotation parameter always has a value between zero and one. For a rotation parameter of zero ($a = 0$) the hole is spherical and does not rotate; for a rotation parameter of 1 ($a = 1$) the hole is highly flattened and rotates extremely rapidly. There is no way to make a hole rotate any faster than $a = 1$; in fact, a hole with a rotation parameter very close to 1 should actually slow down until its rotation parameter is about .998 because of friction with the matter and radiation falling into it.

What are the most important properties that can be deduced from a hole's mass and rotation parameter? First, the gravitational field of the hole obeys the standard laws of Newton and Einstein: the hole's attraction for an object is proportional to its mass and inversely proportional to the square of the distance between it and the object if the distance is somewhat greater than the size of the

hole. Second, a rotating hole creates a vortex in the empty space surrounding it, thereby swirling all particles or gas that approach it into whirlpool orbits. The greater the hole's rotation parameter, the stronger the vortex. Third, a black hole curves space and warps time in its vicinity. Fourth, a black hole has a clearly delineated horizon into which anything can fall but from which nothing can emerge. Fifth, the circumference of a black hole's equator is 19 kilometers multiplied by the mass of the hole and divided by the mass of the sun. Typical black holes should have masses between three and 50 suns, and circumferences between about 60 and 1,000 kilometers.

Such are the predicted properties of black holes from the viewpoint of the theorist. To the observational astronomer these properties present an exciting challenge: find a black hole and verify the predictions! Until the mid-1960's no one took the challenge seriously. Black

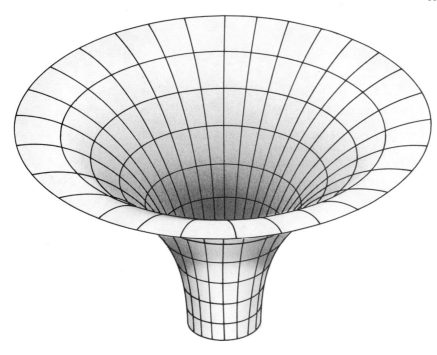

**SPACE IS CURVED in the presence of a strong gravitational field. The curvature is depicted in what is known as an embedding diagram, in which three-dimensional space is represented by a flat plane that is warped by the presence of a star or any other kind of massive object. The amount of curvature is related to the strength of the object's gravitational field, and it affects the direction and travel time of rays of light and the measurement of distances.**

holes were regarded as being strictly theoretical objects: objects that could be formed by the death of a star but probably never were formed. Even if they were, they could probably never be found observationally. Objects such as black holes and neutron stars were too bizarre to fit naturally into our tranquil universe. Somehow all massive stars would eject most of their mass before they died, thereby saving themselves the fate of becoming neutron stars or black holes. This climate of opinion was rarely verbalized explicitly, but it set the tenor of the times. I do not know of a single proposal to search for black holes before 1963.

In the 1960's, however, our view of the universe began to change radically. Exploding galaxies, rapidly varying radio galaxies, quasars, cosmic microwave radiation from the "big bang" explosion that formed the universe, flaring X-ray stars—all these and other observational discoveries taught us how violent and strange the universe can be. Gradually neutron stars and black holes began to seem more plausible. Then, in 1967, pulsars were discovered, and by late 1968 they were shown to be rotating neutron stars beaming radiation out into space. Since neutron stars really existed, then surely black holes must exist as well.

How could one go about searching for a black hole? If black holes are formed by dying massive stars, the nearest black hole should be no closer to the solar system than the nearest massive star: some 10 light-years away. Since most black holes would have a circumference of less than 1,000 kilometers, their resulting angular diameter in the sky would be a millionth of a second of arc. One could certainly not hope to find a black hole as a black spot in the sky.

Could one take advantage of the fact that a black hole's gravitational field can act as a lens and bend and focus light from a more distant star, thereby making the star look temporarily bigger and brighter? This is not a good way to search for black holes. If the hole were close to the star, the amount of focusing would be too small to be noticeable. If the hole and the star were widely separated, the amount of focusing would be large, but interstellar distances are so vast, that the necessary lining up of the earth, the hole and the star would be an exceedingly rare event—so rare that to search for one would be a waste of time. Moreover, even if such an event were observed, it would be impossible to tell whether the gravitational lens was a black hole or merely an ordinary but dim star.

Suppose the black hole had a companion star that could be seen and stud-

ied. Perhaps the presence of the hole could be deduced by its influence on the companion. With this idea in mind two Russian astrophysicists began the first search for black holes in 1964. Ya. B. Zel'dovich and O. Kh. Guseynov looked through catalogues of spectroscopic binary stars for systems that might be a black hole and a normal star revolving around each other. A spectroscopic binary system can look like a single star when it is viewed through even the most powerful telescope. The lines in its spectrum, however, shift periodically from the blue toward the red and back again as the observed star revolves around its darker companion. The periodic shift is produced by the Doppler effect: it is toward the blue as the star moves toward us in its orbit and toward the red as it moves away. The spectral lines of the companion may also be detected, shifting toward the red as the lines from the primary, or brighter, star are shifting toward the blue. In that case the companion is presumably not a black hole. If the companion star cannot be detected, however, it might be a black hole.

Star catalogues are full of binary systems for which the spectral lines of only one star are detected. Several hundred are known, and probably thousands more could be discovered if there were strong reasons to search for them. To shorten the list Zel'dovich and Guseynov investigated the mass of each dark companion. They could estimate the mass roughly from the amount the spectral lines were Doppler-shifted. The more massive the companion, the stronger its pull on the primary star and hence the greater the Doppler shift of the spectral lines. By requiring firm evidence from measurements of the Doppler shift that the mass of the dark companion is three times greater than the mass of the sun (and is therefore not a neutron star or a white dwarf), Zel'dovich and Guseynov brought their list down to a handful of spectroscopic binaries. In some of those systems the primary star was so bright that its dark secondary companion could very well be a normal star masked by the glare of the primary. After discarding those cases Zel'dovich and Guseynov were left with five good candidates for systems incorporating a black hole.

In 1968 Virginia Trimble, who was then working at the California Institute of Technology, and I revised and extended the Zel'dovich-Guseynov list. Unfortunately for us none of the eight good candidates on the new list we prepared presented a truly convincing case for a

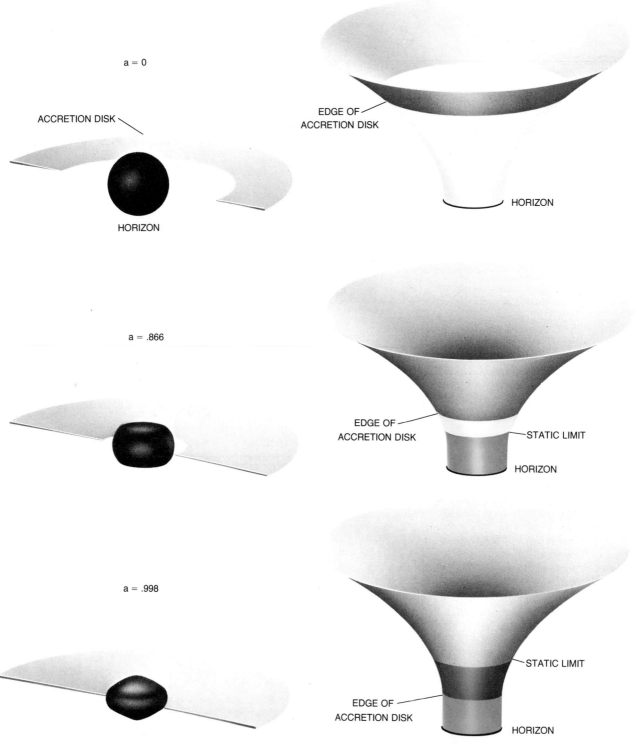

a = 0

ACCRETION DISK

HORIZON

EDGE OF
ACCRETION DISK

HORIZON

a = .866

EDGE OF
ACCRETION DISK

STATIC LIMIT

HORIZON

a = .998

STATIC LIMIT

EDGE OF
ACCRETION DISK

HORIZON

THREE BLACK HOLES, their accretion disks and the curved space that surrounds them are compared to show how they are affected by their speed of rotation. A hole that is not rotating (*top*) has a horizon, or edge, that is spherical. Its rotation parameter $a$ is zero, where the rotation parameter is defined as being equal to the speed of light times the black hole's angular momentum, divided by the Newtonian gravitational constant times the square of the black hole's mass. A hole rotating at a moderate speed (*middle*) with a rotation parameter of .866 will be perfectly flat at the poles but will still be rounded at the equator. Rotation does not affect the size of the equatorial circumference. For a black hole rotating at high speed (*bottom*) with a rotation parameter of .998, the horizon has a shape that cannot exist in the flat Euclidean space of everyday experience. In all three black holes the hot, gaseous ac-

cretion disk that is believed to surround the black hole in Cygnus X-1 is shown. Gas spirals inward through the disk toward the horizon, heating up by friction and emitting X rays along the way. At the inner edge of the disk the hot gas plunges into the hole. A whirlpool motion in space, created by the black hole's rotation, swirls the disk inward so that its inner edge is close to the horizon if the hole is rotating rapidly. The curved space around the hole and the relative positions of the hole and the accretion disk in the curved space are shown by means of an embedding diagram [*see illustration on the preceding page*]. The static limit is the point where the whirlpool motion of space becomes irresistible. For a rapidly rotating black hole the accretion disk extends far below the static limit. X rays emitted from such a disk may have specific characteristics that would reveal whether black-hole theory is correct.

black hole. In all eight cases Trimble was able to conjure up a semireasonable explanation for why the dark companion was invisible without resorting to the hypothesis that it was a black hole. For example, the dim star might itself be a multiple-star system and thus be less luminous than its mass would indicate. Alternatively the primary star might be more luminous than it appeared to be. Or complexities in the spectrum of the primary star might mask the spectral lines of the secondary star. At that point the search for black holes in binary systems seemed to be stymied.

There was one major hope. As early as 1964 it had been realized that a black hole in a close binary system might pull gas off its companion star. As the liberated gas fell into the hole, it might heat up so much that it would emit X rays. Thus if any of the eight good candidates were found to emit X rays, the supposition that the dark companion was a black hole would become much more convincing.

A search for X rays emitted by binary systems could not be conducted with instruments on the ground because the X rays would be absorbed by the earth's atmosphere. One could use instruments aboard sounding rockets. Such a rocket, however, gets only a short peek at the sky before it falls back to the earth, so that its instruments can detect only the brightest of X-ray stars and can examine them only sketchily. Instruments car-

ried aloft by balloons also get only short glimpses, and atmospheric absorption confines these views to the most penetrating of the X rays. For a definitive search an X-ray telescope aboard an artificial satellite would be needed.

The first such telescope was launched jointly by the U.S. and Italy aboard the *Uhuru* satellite on December 12, 1970. By the spring of 1972 *Uhuru* had gathered enough data to compile a detailed catalogue of 125 X-ray sources. To the astronomer searching for black holes the results from *Uhuru* were simultaneously disappointing and encouraging. They were disappointing because none of the X-ray sources coincided with any of the eight black-hole candidates. They were encouraging because at least six of the X-ray sources appeared to lie in other binary systems, typically systems that had not previously been recognized to be binary and had therefore been overlooked in the earlier searches.

Two of the six definite X-ray binary sources, Centaurus X-3 and Hercules X-1, clearly did not harbor a black hole. One could be certain of that because their X rays arrive in precisely timed periodic pulses: 4.84239 seconds between pulses for Centaurus X-3 and 1.23782 seconds for Hercules X-1. Nothing associated with a black hole can give rise to such regular behavior. Presumably each of these two binaries incorporates a rotating neutron star with its

magnetic field inclined to its axis of rotation. Gas that is pulled off the companion star is funneled down the magnetic lines of force onto the magnetic poles of the neutron star, where heat from its impact generates a beam of X rays that sweeps across the sky as the star rotates. Each time the beam sweeps past the earth, the *Uhuru* satellite sees a burst of X rays. A black hole cannot produce such a beam because no off-axis structure such as a magnetic field can ever be anchored in a black hole. The hole would quickly destroy any such structure according to Einstein's laws of gravity.

The four remaining binary X-ray sources are designated 2U 1700-37, 2U 0900-40, SMC X-1 and Cygnus X-1; 2U refers to the second *Uhuru* catalogue and SMC stands for Small Magellanic Cloud, a companion galaxy of our own. Studies at visual wavelengths of each of these sources reveal a supergiant primary star with a telltale periodically varying Doppler shift. There is no sign of spectral lines from the secondary star. In all four cases, however, the visible spectrum shows lines emitted by gas flowing from the primary toward the unseen secondary. The X rays from the three systems 2U 1700-37, 2U 0900-40 and SMC X-1 are eclipsed each time the primary star passes between the earth and the unseen secondary. Therefore the secondary is almost certainly the source of the X rays. The X rays are most likely

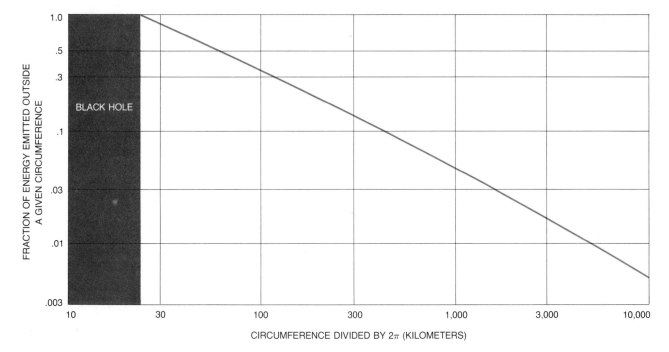

**ENERGY EMITTED** by various regions of the core of the accretion disk around the black hole in Cygnus X-1 is expressed as a fraction of the total energy. Location on the disk is described in units of circumference divided by $2\pi$, which is not equal to radius because space is highly curved in neighborhood of black hole. Half of energy is emitted from innermost 56 kilometers of disk, where temperature is higher than 30 million degrees Kelvin and energy of a typical X-ray photon exceeds 5,000 electron volts.

generated when the flowing gas is heated by falling into the secondary, just as they are when the gas falls on a neutron star; at least astronomers have not been able to invent any other quantitative explanation for the X rays.

To heat the falling gas to the temperatures necessary for the emission of X rays requires huge amounts of energy, energy that can be supplied only by a drop through a very strong gravitational field. The gravitational field surrounding a normal star or planet is not strong enough. Only three types of object have sufficiently strong fields: white dwarfs, neutron stars and black holes. And the only definitive way to distinguish among these three possibilities is to somehow measure the mass of the secondary star. If the mass exceeds 1.4 times the mass of the sun, the object cannot be a white dwarf. If the mass exceeds three times the mass of the sun, the object cannot be a neutron star. In the latter case it must be a black hole.

Obtaining a rough estimate of the mass of the secondary is not too difficult. One needs only a moderate amount of data from observations of the Doppler shift and fairly good information about the spectrum of the primary. On the basis of such data the mass of the unseen companion of 2U 1700-37 is 2.5 times the mass of the sun, the mass of the companion for 2U 0900-40 is three times the mass of the sun, the mass of the

companion for SMC X-1 is twice the mass of the sun and the mass of the companion for Cygnus X-1 is about eight times the mass of the sun.

These figures suggest that at least one and perhaps all four of the X-ray binaries include a black hole. The estimates, however, are quite rough. Individual astronomers interpreting the data in various ways can differ by a factor of two or more in their estimates of masses. And any astronomer who wants to play the devil's advocate can reduce the estimates still further by introducing peculiar interpretations of the data.

Only in the case of Cygnus X-1 do the devil's advocates face difficulties. Even a "worst-case" analysis of these data reveals that the unseen companion has a mass of no less than four times the mass of the sun. Therefore one can conclude that Cygnus X-1 does comprise a black hole. At least this conclusion is the most reasonable one.

Teams of devil's advocates led by John N. Bahcall of Princeton University and James Pringle of the University of Cambridge have invented viable, although less plausible, alternative models to explain the observations. The models assume that the massive secondary around which the bright primary travels is a normal but dim star. In one model the X rays come from a satellite neutron star in orbit around a massive normal secondary. In another model a neutron star emitting X rays circles in a wide or-

bit around an entire normal binary system. In a third model the X rays do not come from a neutron star or any other compact object at all. Instead between two normal stars stretch strong magnetic fields that are continually being twisted, tied into knots and broken as the two stars rotate. The knotting and breaking of the fields heats gas that is attached to them, and the hot gas emits the X rays. A fourth model assumes that the secondary emitting the X rays is not a black hole but is something even more exciting and bizarre: a massive "naked singularity" in the structure of space-time. Einstein's laws of gravity probably forbid the formation of naked singularities, but all attempts to prove that this is the case have failed.

We now face a situation that is common in astronomy. One model, that a black hole is in orbit around a normal star, can readily explain the observations: the light and the X rays from Cygnus X-1. The other leading models, proposing a secondary that is either a white dwarf or a neutron star, have been killed by the observations. Alternative contrived explanations, however, can still be made to fit the data. Such a situation leads astronomers into the final stage of the search for black holes: the attempt to accumulate more data of higher quality and of new types, data that (I hope!) will gradually kill the contrived models and clinch the case for a

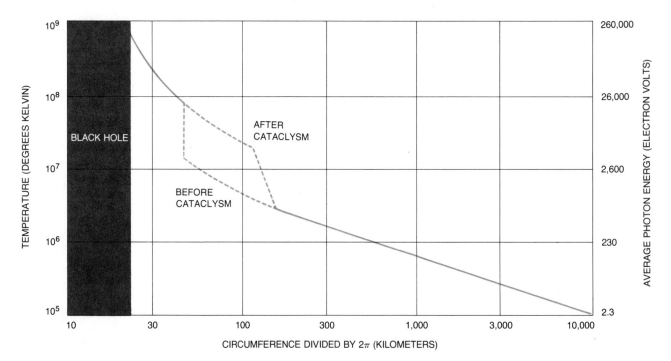

**TEMPERATURE OF GAS AND ENERGY OF PHOTONS emitted** by core of the accretion disk is shown for various locations. Only X rays more energetic than 2,000 electron volts have been detected with X-ray telescopes available until recently; thus the curve for the less energetic photons is calculated from theory. Future X-ray telescopes will test theory by examining photons between 400 and 2,000 electron volts. Cataclysmic change in Cygnus X-1 may have been a change of state in the region between 50 and 100 kilometers.

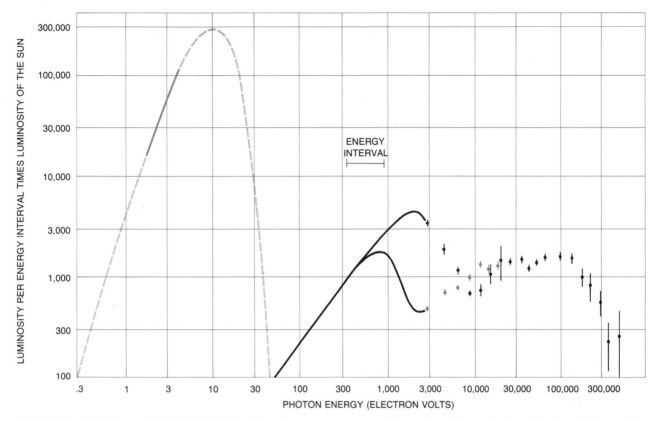

SPECTRUM OF HDE 226868 AND CYGNUS X-1 is shown in terms of the absolute luminosity of the sun. The unit energy interval is the interval over which the energy increases by a factor of the base of natural logarithms, e, or about 2.718, for example at an energy between 1,000 electron volts and 2,718 electron volts. The curve in color shows the spectrum of HDE 226868; the solid portion of the curve is known from visual observations and the broken portion is extrapolated into the infrared and ultraviolet regions of the spectrum. The calculated emission from the core of the accretion disk of Cygnus X-1 is shown in gray. The points at energies above 2,700 electron volts are X-ray observations of Cygnus X-1 averaged over short fluctuations ranging from milliseconds to minutes; the vertical error bar on each point shows the statistical uncertainty due to the fact that the observations consisted of only a finite number of photons. Black dots are observations made before the cataclysmic change; colored dots show how the spectrum was altered.

black hole. In this final stage of the search it will be helpful to know what to look for. What kind of signatures should characterize the X rays and the light generated by a binary system whose secondary is a black hole accreting gas from a primary star?

Detailed theoretical studies of such binary systems were begun in 1971 before the observations from *Uhuru* had confirmed the fact that some binaries do emit X rays. The studies were initiated independently on the basis of Newtonian gravitation theory by Nikolai Shakura and Rashid Sunyaev of the Institute of Applied Mathematics in Moscow and by Pringle and Martin J. Rees at the University of Cambridge. Later analyses based on Einstein's general theory of relativity were undertaken by Donald Page and me at the California Institute of Technology, by Igor Novikov and Andrei Polnarev of the Institute of Applied Mathematics in Moscow and by Christopher Cunningham at the University of Washington. All of these studies reveal the same gross structure for the binary system and the flow of gas within it [*see illustration on page 64*].

The gravitational field of the black hole is continually pulling gas off of the supergiant star or out of its immediate vicinity and funneling it into orbits around the black hole. Centrifugal and gravitational forces flatten the orbiting gas into a thin disk around the black hole that is analogous to the rings around Saturn but is far larger. Once a gas filament is sucked into the disk, it would stay in orbit around the hole forever if there were no friction. Friction between adjacent gas filaments, however, forces the gas to spiral slowly inward. The inward velocity is much less than 1 percent of the orbital velocity. A few weeks or months after a filament enters the disk it has spiraled inward by several million kilometers and is approaching the disk's inner edge. There the black hole's gravity becomes irresistible. It sucks the gas filament away from the disk and, within a fraction of a second, through the black hole's horizon into the hole itself.

A by-product of the friction is heat.

When a filament is first caught in the disk it might have a temperature of 25,000 degrees Kelvin, the same temperature as the surface of the supergiant primary. As the gas drifts inward through the disk friction heats it until in the last 100 kilometers of its spiraling descent it is hotter than 10 million degrees.

The hot gas radiates energy, about 80 percent of which is emitted from the inner 200 kilometers as X rays. Presumably these are the X rays detected by the telescope aboard *Uhuru*. The remaining 20 percent of the radiation, emitted from the comparatively cool outer parts of the disk, should be less energetic X rays that cannot be detected with the instruments currently available, together with ultraviolet radiation and light that would not be detectable against the glare of the supergiant star.

The case for a black hole in Cygnus X-1 would be much strengthened if theorists working with this model could calculate the properties of the X-ray emission in detail. The chief stumbling block at this point is friction in the disk. We do not know whether the friction is generated by turbulence in the spiraling gas, by magnetic fields embedded in the spiraling gas or by a combination of turbulence and magnetic fields. Even if we did know the source of the friction, we could not calculate its magnitude because we do not yet know enough about the general physical behavior of turbulent magnetized gases.

It is remarkable that in spite of our ignorance we are still able to calculate with confidence some of the important features of the disk. For example, from the laws of the conservation of energy and of angular momentum we can calculate how much energy each region of the disk radiates. We have concluded that most of the radiation must come from the hot inner 200 kilometers, no matter what the source or magnitude of the friction may be. We cannot, however, calculate the temperature in that inner region or the spectrum of the X rays it should emit. Instead we must discover what the spectrum is from observations and from it infer that the temperatures range between five million and 500 million degrees K. We go back to the models and see that such high temperatures are incompatible with a calm disk having little friction or turbulence but are quite reasonable if the inner region of the disk is violently turbulent and is optically

thin, that is, translucent to radiation. We then return to the observations and note that the X rays do not arrive at the earth steadily. Their intensity fluctuates by a factor of two or three or even more over any length of time from milliseconds to days. Such fluctuations are also what one might expect from a turbulent disk.

By working back and forth between theory and observations in this way Richard Price of the University of Utah and I have built up a workable description of the structure of the inner disk of Cygnus X-1. Of course, success in building such a model is not much of a positive addition to the explanation that Cygnus X-1 includes a black hole. Too much was inferred from the observations and too little was calculated from the basic assumptions or the first principles of physics. On the other hand, things could be worse. The observations might have been incompatible with any type of model that assumed that the X rays were emitted by gas flowing into a black hole. In a sense, then, the black-hole model has survived a negative test. This type of test is the chief tool by which astrophysicists prove and disprove models. When a model has withstood a variety of negative tests that have destroyed all its competitors, astrophysicists begin to take the model very seriously.

Negative tests may not be the only way to prove or disprove the black-hole

explanation of Cygnus X-1. Sunyaev has suggested one positive test and other theorists are searching for more. Sunyaev's test consists in looking for brief flares in the intensity of the X rays. Such X-ray flares would presumably be generated by temporary hot spots in the inner 100 kilometers of the disk [see top illustration, p. 74]. We have no reliable theory that accounts for the origin and destruction of such hot spots. Nevertheless it seems likely that once a hot spot is born it would live for more than one circuit in its orbit around the black hole. The X rays from the hot spot would be beamed in a direction that rotates as the hot spot circles the black hole. The beaming might be caused in part by the process by which the radiation is emitted and in part by focusing of the radiation in the gravitational field of the black hole. Moreover, the Doppler shift would make the X rays more intense as the hot spot approached us in its orbit around the hole, and less intense as it receded. Hence the emission from the hot spot would not arrive steadily but should arrive in bursts, with the interval between bursts equal to the orbital period of the hot spot traveling around the black hole. Thus the model predicts that short X-ray flares are likely to show a substructure of pulses with an interval between pulses of a few milliseconds.

If such pulses were perceived and if the black hole's mass were known, then

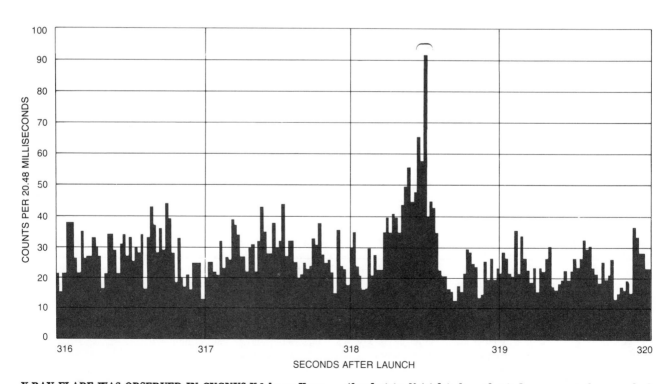

X-RAY FLARE WAS OBSERVED IN CYGNUS X-1 by an X-ray telescope aboard a rocket built by the Goddard Space Flight Center and launched in October, 1973 (*left*). A closer analysis of the flare (*bracket*) itself (*right*) shows that it does appear to have a pulsed substructure (*color*) near the telescope's limit of sensitivity. (Apparent pulses in black are random fluctuations in the X-ray signal.)

the interval between pulses would provide a way of computing the circumference of the hot spot's orbit around the hole. By observing many pulsed X-ray flares and determining the minimum interval between pulses we could learn what the circumference of the inner edge of the disk is, and from that we could infer the speed at which the hole is rotating. For a black hole with a mass eight times the mass of the sun the minimum interval between pulses must be between 3.6 milliseconds for a nonrotating hole (rotation parameter $a = 0$) and .6 millisecond for a rapidly rotating hole (rotation parameter $a = .998$).

Such a pulsed substructure is not, however, obligatory for X-ray flares. Flares without a pulsed substructure could originate in hot spots that are far enough from the black hole (perhaps more than 100 kilometers) so that they are not beamed by the hole's gravitational field and do not vary much because of the Doppler shift. They could also originate in spots that are very large and quickly get strung out into a doughnut around the disk by the orbital motion of the gas. Thus the observation of nonpulsed flares is also quite compatible with the black-hole explanation of Cygnus X-1. If pulsed X-ray flares can be detected, however, they should pulse only in the range predicted by Sunyaev's arguments.

The search for pulsed flares calls for an X-ray telescope that can make an observation in less than one millisecond and that has more than 10,000 square centimeters of area for collecting X-ray photons, so that many photons can be counted in each millisecond. Such a telescope will probably not be put into orbit around the earth until the National Aeronautics and Space Administration launches its first High Energy Astronomical Observatory satellite (HEAO-A), perhaps in 1977. Between now and then we must content ourselves with brief glimpses from the instruments on rockets and balloons.

The first such glimpse, in October, 1973, was promising. X-ray flares lasting about .1 second were observed with a telescope having a time resolution of .32 millisecond and the modest collecting area of 1,360 square centimeters. The largest of the flares does appear to have a pulsed substructure, but not enough photons were counted in the flare to be certain. So here we theorists sit, impatiently awaiting the next generation of X-ray telescopes, those of us in the "establishment" trying with great difficulty to build better black-hole models of Cygnus X-1 and those of us who are devil's advocates trying equally hard to build better non-black-hole models.

While some of us struggle with Cygnus X-1, others search elsewhere for black holes. The possibilities are plentiful, but none has yet yielded strong evidence for a hole. Several other spectroscopic binaries, including Epsilon Aurigae and Beta Lyrae, include a secondary star that has a mass of more than four times the mass of the sun and is surrounded by a huge opaque, or partially opaque, disk. As seen from the earth, the disk periodically blots out the light from the bright primary star. A. G. W. Cameron of the Harvard College Observatory and Edward Devinney of the University of South Florida have suggested that massive objects at the center of these disks might be black holes. Other astronomers find other explanations equally plausible.

The Russian astronomer V. F. Schwartzman is searching for black holes that have no binary companion. He calculates that the interstellar gas being sucked into such an isolated hole should emit light that flickers with a period of several milliseconds. Unfortunately for observers the light would not be very intense; it would have no more than 1 percent of the intensity of the sun's light if the sun were being observed at that distance. Since the nearest such black hole would be many light-years away, it would appear as a faint, rapidly flickering star. To detect the flicker and thereby make a strong case for the existence of an isolated black hole, Schwartzman needs sensitive electronics and a powerful telescope for observing at visible wavelengths. He is working at the Crimean Astrophysical Observatory, where a 240-inch telescope is nearing completion.

The collapse of the star that gives rise to a black hole should also generate a huge burst of gravitational waves. The present first-generation gravitational-wave antennas are only sensitive enough to detect such bursts from our own galaxy, where they would not be expected more than once every few years. Second-generation antennas now under construction at Stanford University, Louisiana State University and the University of Moscow might be able to detect such bursts from the cluster of 2,500 galaxies in the constellation Virgo. Even more sensitive third-generation antennas will surely be able to do so. Theorists expect that in the Virgo cluster there will be several black holes born every year. By detecting and analyzing the gravitational waves from such births one could not only verify the creation of a black hole but also study some intimate details of the newborn hole. It is a project a decade or so in the future.

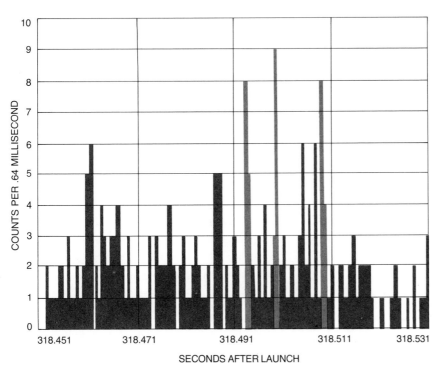

The time interval between pulses was no shorter than .005 second, corresponding to a circumference no smaller than four times the circumference of the black hole. That circumference is not small enough to determine the rotation parameter of the black hole.

Thus far I have described only normal black holes created by the collapse of normal stars ranging in mass from three to 60 times the mass of the sun. There are probably supermassive holes and possibly miniholes as well. Donald Lyn-den-Bell of the University of Cambridge has argued that the dense milieu of gas and stars that fuels grand-scale explosions in the nuclei of some galaxies must ultimately collapse to form a supermassive black hole. If this is true, a galaxy such as our own, which probably had explosions in its nucleus long ago, might possess a huge black hole in its nucleus today. That hole would be a "tomb" from a more violent past. Such a black hole in our own galaxy might be as massive as 100 million times the mass of the sun and have a circumference as large as two billion kilometers. The hole would suck gas from the surrounding galactic nucleus, perhaps forming a gigantic accretion disk analogous to the disks proposed for the spectroscopic-binary systems. Lynden-Bell and Rees calculate that such a disk would emit strong radio and infrared radiation but not X rays. The nucleus of our galaxy does give evidence of several bright infrared and radio "stars." Unfortunately for theorists an accretion disk around a supermassive black hole is not the only possible explanation of the observed objects, and so far no one has invented a definitive test for the hypothesis.

Miniholes far less massive than the sun cannot be created in the universe as it exists today. Nature simply does not supply the necessary compressional forces. The necessary forces were present, however, in the first few moments after the creation of the universe in the "big bang." If the big bang were sufficiently chaotic, then, according to calculations made by Hawking, it should have produced a great number of miniholes. Hawking has shown that miniholes behave quite differently from normal-sized holes. Any hole less massive than $10^{16}$ grams (the mass of a small iceberg) should gradually destroy itself by an emission of light and particles according to certain laws of quantum mechanics. Those laws, which are not important for the larger black holes, considerably modify the properties of the smaller holes. The result is that all the primordial black holes less massive than $10^{15}$ grams should be gone by now. Those with a mass between $10^{15}$ grams and $10^{16}$ grams are now dying. In its final death throes such a dying black hole would not be black at all. It would be a fireball powerful enough to supply all the energy needs of the earth for several decades yet small enough to fit inside the nucleus of an atom.

Hawking's results are less than a year old, and so their implications have not yet been explored in detail. They may motivate a flood of proposals for searching for miniholes. He and Page are exploring one possibility: that the bursts of cosmic gamma rays that have been detected by instruments on artificial satellites of the Vela series came from explosions of miniholes.

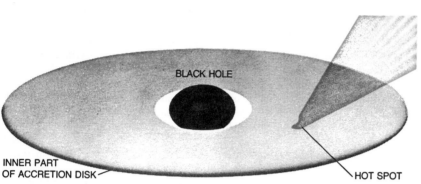

RADIATION FROM A HOT SPOT on the accretion disk of a black hole would be beamed into a wide cone that would sweep out a circle in the sky above the disk each time the spot made an orbit around the hole. As the cone sweeps repetitively past the earth X-ray telescopes would observe pulses of X rays amidst a general increase in the total intensity of the X rays received. The result would be a flare of X rays that would have a pulsed substructure.

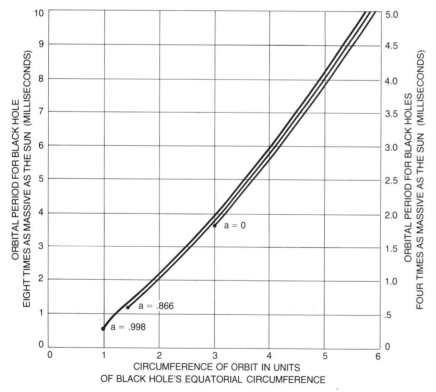

TIME INTERVAL BETWEEN PULSES emitted by a hot spot in orbit on the accretion disk around a black hole is uniquely and precisely determined by the circumference of the spot's orbit and by the mass and rotation parameter a of the hole. The more rapidly the hole rotates, the smaller the inner edge of its accretion disk is, and hence the shorter the minimum time would be between pulses in an X-ray flare. Observations of such pulses in X-ray flares can provide a way of measuring the rotation parameter of a black hole in Cygnus X-1.

The present list of ways and places that black holes might be found is far from complete. With so many possibilities a theorist such as the author cannot help being excited—until he talks with his more down-to-earth experimenter friends. Then he realizes what a difficult job the search really is. We cannot expect quick results, but the future does not seem unpromising.

# 9

# The Quantum Mechanics
# of Black Holes

by S. W. Hawking
*January 1977*

*Black holes are often defined as areas from which nothing, not even light, can escape. There is good reason to believe, however, that particles can get out of them by "tunneling"*

The first 30 years of this century saw the emergence of three theories that radically altered man's view of physics and of reality itself. Physicists are still trying to explore their implications and to fit them together. The three theories were the special theory of relativity (1905), the general theory of relativity (1915) and the theory of quantum mechanics (c. 1926). Albert Einstein was largely responsible for the first, was entirely responsible for the second and played a major role in the development of the third. Yet Einstein never accepted quantum mechanics because of its element of chance and uncertainty. His feelings were summed up in his often-quoted statement "God does not play dice." Most physicists, however, readily accepted both special relativity and quantum mechanics because they described effects that could be directly observed. General relativity, on the other hand, was largely ignored because it seemed too complicated mathematically, was not testable in the laboratory and was a purely classical theory that did not seem compatible with quantum mechanics. Thus general relativity remained in the doldrums for nearly 50 years.

The great extension of astronomical observations that began early in the 1960's brought about a revival of interest in the classical theory of general relativity because it seemed that many of the new phenomena that were being discovered, such as quasars, pulsars and compact X-ray sources, indicated the existence of very strong gravitational fields, fields that could be described only by general relativity. Quasars are starlike objects that must be many times brighter than entire galaxies if they are as distant as the reddening of their spectra indicates; pulsars are the rapidly blinking remnants of supernova explosions, believed to be ultradense neutron stars; compact X-ray sources, revealed by instruments aboard space vehicles, may also be neutron stars or may be hypothetical objects of still higher density, namely black holes.

One of the problems facing physicists who sought to apply general relativity to these newly discovered or hypothetical objects was to make it compatible with quantum mechanics. Within the past few years there have been developments that give rise to the hope that before too long we shall have a fully consistent quantum theory of gravity, one that will agree with general relativity for macroscopic objects and will, one hopes, be free of the mathematical infinities that have long bedeviled other quantum field theories. These developments have to do with certain recently discovered quantum effects associated with black holes, which provide a remarkable connection between black holes and the laws of thermodynamics.

Let me describe briefly how a black hole might be created. Imagine a star with a mass 10 times that of the sun. During most of its lifetime of about a billion years the star will generate heat at its center by converting hydrogen into helium. The energy released will create sufficient pressure to support the star against its own gravity, giving rise to an object with a radius about five times the radius of the sun. The escape velocity from the surface of such a star would be about 1,000 kilometers per second. That is to say, an object fired vertically upward from the surface of the star with a velocity of less than 1,000 kilometers per second would be dragged back by the gravitational field of the star and would return to the surface, whereas an object with a velocity greater than that would escape to infinity.

When the star had exhausted its nuclear fuel, there would be nothing to maintain the outward pressure, and the star would begin to collapse because of its own gravity. As the star shrank, the gravitational field at the surface would become stronger and the escape velocity would increase. By the time the radius had got down to 30 kilometers the escape velocity would have increased to 300,000 kilometers per second, the velocity of light. After that time any light emitted from the star would not be able to escape to infinity but would be dragged back by the gravitational field. According to the special theory of relativity nothing can travel faster than light, so that if light cannot escape, nothing else can either.

The result would be a black hole: a region of space-time from which it is not possible to escape to infinity. The boundary of the black hole is called the event horizon. It corresponds to a wave front of light from the star that just fails to escape to infinity but remains hovering at the Schwarzschild radius: $2GM/c^2$, where $G$ is Newton's constant of gravity, $M$ is the mass of the star and $c$ is the velocity of light. For a star of about 10 solar masses the Schwarzschild radius is about 30 kilometers.

There is now fairly good observational evidence to suggest that black holes of about this size exist in double-star systems such as the X-ray source known as Cygnus X-1 [see the article "The Search for Black Holes," by Kip S. Thorne, beginning on page 63]. There might also be quite a number of very much smaller black holes scattered around the universe, formed not by the collapse of stars but by the collapse of highly compressed regions in the hot, dense medium that is believed to have existed shortly after the "big bang" in which the universe originated. Such "primordial" black holes are of greatest interest for the quantum effects I shall describe here. A black hole weighing a billion tons (about the mass of a mountain) would have a radius of about $10^{-13}$ centimeter (the size of a neutron or a proton). It could be in orbit either around the sun or around the center of the galaxy.

The first hint that there might be a connection between black holes and thermodynamics came with the mathematical discovery in 1970 that the surface area of the event horizon, the boundary of a black hole, has the property that it always increases when addi-

tional matter or radiation falls into the black hole. Moreover, if two black holes collide and merge to form a single black hole, the area of the event horizon around the resulting black hole is greater than the sum of the areas of the event horizons around the original black holes. These properties suggest that there is a resemblance between the area of the event horizon of a black hole and

the concept of entropy in thermodynamics. Entropy can be regarded as a measure of the disorder of a system or, equivalently, as a lack of knowledge of its precise state. The famous second law of thermodynamics says that entropy always increases with time.

The analogy between the properties of black holes and the laws of thermodynamics has been extended by James M.

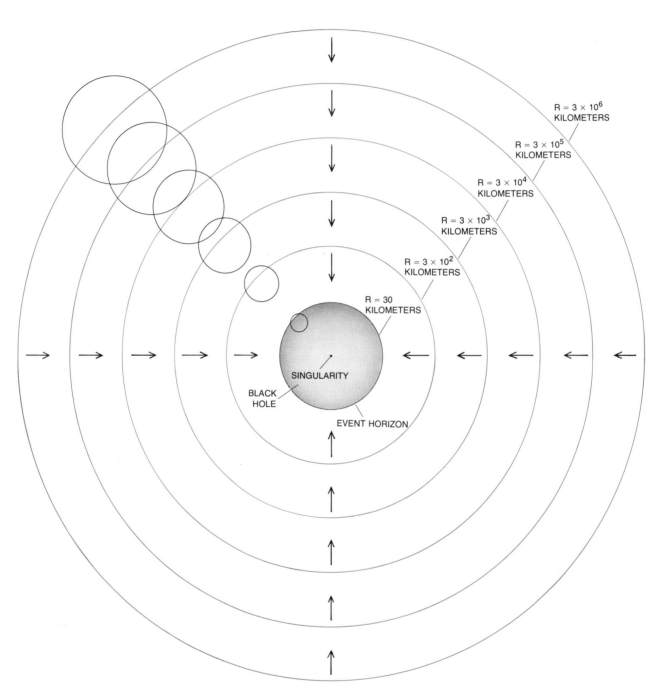

**COLLAPSE OF STAR** of 10 solar masses is depicted schematically from an original radius of three million kilometers (about five times the radius of the sun) to 30 kilometers, when it disappears within the "event horizon" that defines the outer limits of a black hole. The star continues to collapse to what is called a space-time singularity, about which the laws of physics are silent. The series of six small circles represents the wave fronts of light emitted from the successive sur-

faces an instant before the star had collapsed to the dimensions shown. Radii of the star and of the wave fronts are on a logarithmic scale. At each stage of collapse more of the wave front falls within the volume of the star as the escape velocity increases from 1,000 kilometers per second to 300,000 kilometers per second, the velocity of light. The final velocity is reached as the star disappears within the event horizon. No light emitted after that can ever reach outside observers.

Bardeen of the University of Washington, Brandon Carter, who is now at the Meudon Observatory, and me. The first law of thermodynamics says that a small change in the entropy of a system is accompanied by a proportional change in the energy of the system. The fact of proportionality is called the temperature of the system. Bardeen, Carter and I found a similar law relating the change in mass of a black hole to a change in the area of the event horizon. Here the factor of proportionality involves a quantity called the surface gravity, which is a measure of the strength of the gravitational field at the event horizon. If one accepts that the area of the event horizon is analogous to entropy, then it would seem that the surface gravity is analogous to temperature. The resemblance is strengthened by the fact that the surface gravity turns out to be the same at all points on the event horizon, just as the temperature is the same everywhere in a body at thermal equilibrium.

Although there is clearly a similarity between entropy and the area of the event horizon, it was not obvious to us how the area could be identified as the entropy of a black hole. What would be meant by the entropy of a black hole? The crucial suggestion was made in 1972 by Jacob D. Bekenstein, who was then a graduate student at Princeton University and is now at the University of the Negev in Israel. It goes like this. When a black hole is created by gravitational collapse, it rapidly settles down to a stationary state that is characterized by only three parameters: the mass, the angular momentum and the electric charge. Apart from these three properties the black hole preserves no other details of the object that collapsed. This conclusion, known as the theorem "A black hole has no hair," was proved by the combined work of Carter, Werner Israel of the University of Alberta, David C. Robinson of King's College, London, and me.

The no-hair theorem implies that a large amount of information is lost in a gravitational collapse. For example, the final black-hole state is independent of whether the body that collapsed was composed of matter or antimatter and whether it was spherical or highly irregular in shape. In other words, a black hole of a given mass, angular momentum and electric charge could have been formed by the collapse of any one of a large number of different configurations of matter. Indeed, if quantum effects are neglected, the number of configurations would be infinite, since the black hole could have been formed by the collapse of a cloud of an indefinitely large number of particles of indefinitely low mass.

The uncertainty principle of quantum

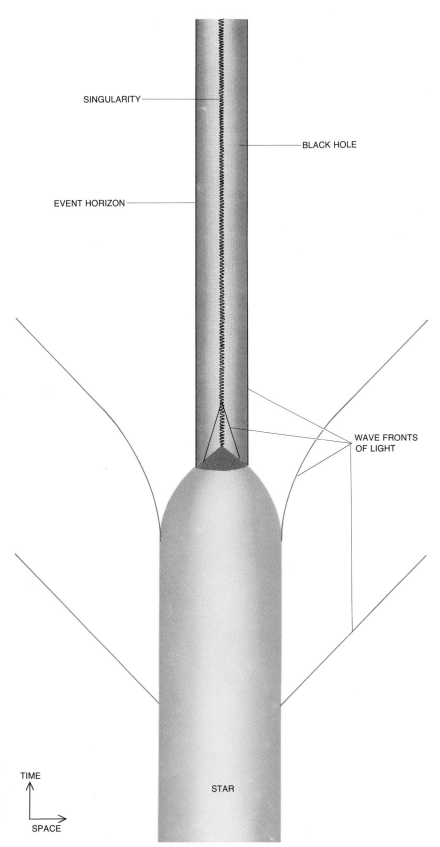

**GRAVITATIONAL COLLAPSE OF A STAR** is depicted in a space-time diagram in which two of the three dimensions of space have been suppressed. The vertical dimension is time. When the radius of the star reaches a critical value, the Schwarzschild radius, the light emitted by the star can no longer escape but remains hovering at that radius, forming the event horizon, the boundary of the black hole. Inside black hole star continues collapse to a singularity.

mechanics implies, however, that a particle of mass $m$ behaves like a wave of wavelength $h/mc$, where $h$ is Planck's constant (the small number $6.62 \times 10^{-27}$ erg-second) and $c$ is the velocity of light. In order for a cloud of particles to be able to collapse to form a black hole it would seem necessary for this wavelength to be smaller than the size of the black hole that would be formed. It therefore appears that the number of configurations that could form a black hole of a given mass, angular momentum and electric charge, although very large, may be finite. Bekenstein suggested that one could interpret the logarithm of this number as the entropy of a black hole. The logarithm of the number would be a measure of the amount of information that was irretrievably lost during the collapse through the event horizon when a black hole was created.

The apparently fatal flaw in Bekenstein's suggestion was that if a black hole has a finite entropy that is proportional to the area of its event horizon, it also ought to have a finite temperature, which would be proportional to its surface gravity. This would imply that a black hole could be in equilibrium with thermal radiation at some temperature other than zero. Yet according to classical concepts no such equilibrium is possible, since the black hole would absorb any thermal radiation that fell on it but by definition would not be able to emit anything in return.

This paradox remained until early in 1974, when I was investigating what the behavior of matter in the vicinity of a black hole would be according to quantum mechanics. To my great surprise I found that the black hole seemed to emit particles at a steady rate. Like everyone else at that time, I accepted the dictum that a black hole could not emit anything. I therefore put quite a lot of effort into trying to get rid of this embarrassing effect. It refused to go away, so that in the end I had to accept it. What finally convinced me it was a real physical process was that the outgoing particles have a spectrum that is precisely thermal: the black hole creates and emits particles and radiation just as if it were an ordinary hot body with a temperature that is proportional to the surface gravity and inversely proportional to the mass. This made Bekenstein's suggestion that a black hole had a finite entropy fully consistent, since it implied that a black hole could be in thermal equilibrium at some finite temperature other than zero.

Since that time the mathematical evidence that black holes can emit thermally has been confirmed by a number of other people with various different approaches. One way to understand the emission is as follows. Quantum mechanics implies that the whole of space is filled with pairs of "virtual" particles and antiparticles that are constantly materializing in pairs, separating and then coming together again and annihilating each other. These particles are called virtual because, unlike "real" particles, they cannot be observed directly with a particle detector. Their indirect effects can nonetheless be measured, and their existence has been confirmed by a small shift (the "Lamb shift") they produce in the spectrum of light from excited hydrogen atoms. Now, in the presence of a black hole one member of a pair of virtual particles may fall into the hole, leaving the other member without a partner with which to annihilate. The forsaken particle or antiparticle may fall into the black hole after its partner, but it may also escape to infinity, where it appears to be radiation emitted by the black hole.

Another way of looking at the process is to regard the member of the pair of particles that falls into the black hole—the antiparticle, say—as being really a particle that is traveling backward in time. Thus the antiparticle falling into the black hole can be regarded as a particle coming out of the black hole but traveling backward in time. When the particle reaches the point at which the

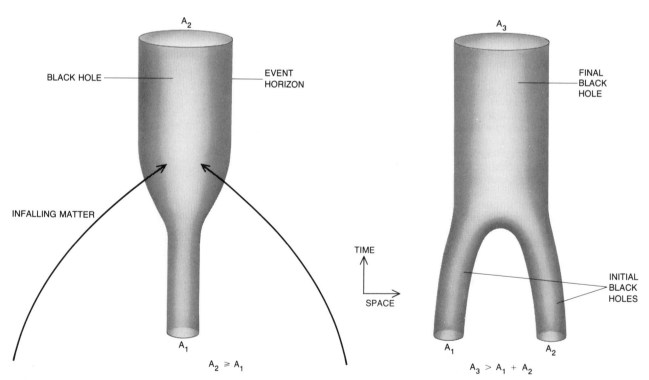

CERTAIN PROPERTIES OF BLACK HOLES suggest that there is a resemblance between the area of the event horizon of a black hole and the concept of entropy in thermodynamics. As matter and radiation continue to fall into a black hole (space-time configuration at left) the area of the cross section of the event horizon steadily increases. If two black holes collide and merge (configuration at right), the area of the cross section of the event horizon of the resulting black hole is greater than the sum of the areas of the event horizons of the initial black holes. The second law of thermodynamics says that the entropy of an isolated system always increases with passage of time.

particle-antiparticle pair originally materialized, it is scattered by the gravitational field so that it travels forward in time.

Quantum mechanics has therefore allowed a particle to escape from inside a black hole, something that is not allowed in classical mechanics. There are, however, many other situations in atomic and nuclear physics where there is some kind of barrier that particles should not be able to penetrate on classical principles but that they are able to tunnel through on quantum-mechanical principles.

The thickness of the barrier around a black hole is proportional to the size of the black hole. This means that very few particles can escape from a black hole as large as the one hypothesized to exist in Cygnus X-1 but that particles can leak very rapidly out of smaller black holes. Detailed calculations show that the emitted particles have a thermal spectrum corresponding to a temperature that increases rapidly as the mass of the black hole decreases. For a black hole with the mass of the sun the temperature is only about a ten-millionth of a degree above absolute zero. The thermal radiation leaving a black hole with that temperature would be completely swamped by the general background of radiation in the universe. On the other hand, a black hole with a mass of only a billion tons, that is, a primordial black hole roughly the size of a proton, would have a temperature of some 120 billion degrees Kelvin, which corresponds to an energy of some 10 million electron volts. At such a temperature a black hole would be able to create electron-positron pairs and particles of zero mass, such as photons, neutrinos and gravitons (the presumed carriers of gravitational energy). A primordial black hole would release energy at the rate of 6,000 megawatts, equivalent to the output of six large nuclear power plants.

As a black hole emits particles its mass and size steadily decrease. This makes it easier for more particles to tunnel out, and so the emission will continue at an ever increasing rate until eventually the black hole radiates itself out of existence. In the long run every black hole in the universe will evaporate in this way. For large black holes, however, the time it will take is very long indeed: a black hole with the mass of the sun will last for about $10^{66}$ years. On the other hand, a primordial black hole should have almost completely evaporated in the 10 billion years that have elapsed since the big bang, the beginning of the universe as we know it. Such black holes should now be emitting hard gamma rays with an energy of about 100 million electron volts.

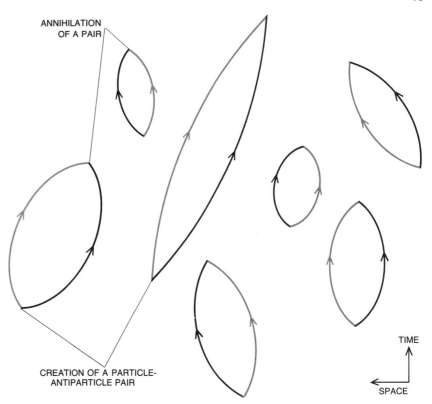

**"EMPTY" SPACE-TIME is full of "virtual" pairs of particles (*black*) and antiparticles (*color*). Members of a pair come into existence simultaneously at a point in space-time, move apart and come together again, annihilating each other. They are called virtual because unlike "real" particles they cannot be detected directly. Their indirect effects can nonetheless be measured.**

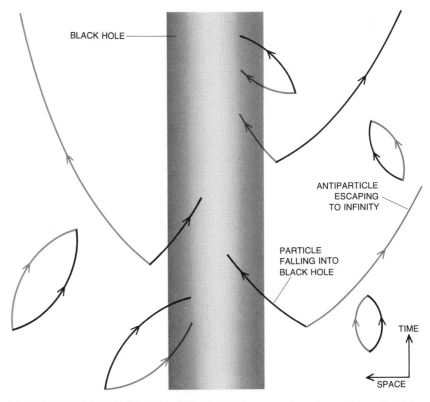

**IN THE NEIGHBORHOOD OF A BLACK HOLE one member of a particle-antiparticle pair may fall into the black hole, leaving the other member of the pair without a partner with which to annihilate. If surviving member of pair does not follow its partner into black hole, it may escape to infinity. Thus black hole will appear to be emitting particles and antiparticles.**

Calculations made by Don N. Page of the California Institute of Technology and me, based on measurements of the cosmic background of gamma radiation made by the satellite SAS-2, show that the average density of primordial black holes in the universe must be less than about 200 per cubic light-year. The local density in our galaxy could be a million times higher than this figure if primordial black holes were concentrated in the "halo" of galaxies—the thin cloud of rapidly moving stars in which each galaxy is embedded—rather than being uniformly distributed throughout the universe. This would imply that the primordial black hole closest to the earth is probably at least as far away as the planet Pluto.

The final stage of the evaporation of a black hole would proceed so rapidly that it would end in a tremendous explosion. How powerful this explosion would be depends on how many different species of elementary particles there are. If, as is now widely believed, all particles are made up of perhaps six different kinds of quarks, the final explosion would have an energy equivalent to about 10 million one-megaton hydrogen bombs. On the other hand, an alternative theory of elementary particles put forward by R. Hagedorn of the European Organization for Nuclear Research argues that there is an infinite number of elementary particles of higher and higher mass. As a black hole got smaller and hotter, it would emit a larger and larger number of different species of particles and would produce an explosion perhaps 100,000 times more powerful than the one calculated on the quark hypothesis. Hence the observation of a black-hole explosion would provide very important information on elementary particle physics, information that might not be available any other way.

A black-hole explosion would produce a massive outpouring of high-energy gamma rays. Although they might be observed by gamma-ray detectors on satellites or balloons, it would be difficult to fly a detector large enough to have a reasonable chance of intercepting a significant number of gamma-ray photons from one explosion. One possibility would be to employ a space shuttle to build a large gamma-ray detector in orbit. An easier and much cheaper alternative would be to let the earth's upper atmosphere serve as a detector. A high-energy gamma ray plunging into the atmosphere will create a shower of electron-positron pairs, which initially will be traveling through the atmosphere faster than light can. (Light is slowed down by interactions with the air molecules.) Thus the electrons and positrons will set up a kind of sonic boom, or shock wave, in the electromagnetic field. Such a shock wave, called Cerenkov ra-

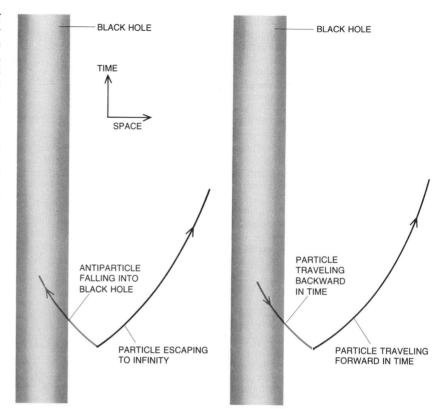

**ALTERNATIVE INTERPRETATIONS** can explain the emission of particles by a black hole. One explanation (*left*) invokes the formation of a virtual particle-antiparticle pair, one member of which is trapped by the black hole as the other escapes. In another explanation (*right*) one can regard an antiparticle falling into a black hole as being a normal particle that is traveling backward in time out of the black hole. Once outside it is scattered by the gravitational field and converted into a particle traveling forward in time, which escapes to infinity.

diation, could be detected from the ground as a flash of visible light.

A preliminary experiment by Neil A. Porter and Trevor C. Weekes of University College, Dublin, indicates that if black holes explode the way Hagedorn's theory predicts, there are fewer than two black-hole explosions per cubic light-year per century in our region of the galaxy. This would imply that the density of primordial black holes is less than 100 million per cubic light-year. It should be possible to greatly increase the sensitivity of such observations. Even if they do not yield any positive evidence of primordial black holes, they will be very valuable. By placing a low upper limit on the density of such black holes, the observations will indicate that the early universe must have been very smooth and nonturbulent.

The big bang resembles a black-hole explosion but on a vastly larger scale. One therefore hopes that an understanding of how black holes create particles will lead to a similar understanding of how the big bang created everything in the universe. In a black hole matter collapses and is lost forever but new matter is created in its place. It

may therefore be that there was an earlier phase of the universe in which matter collapsed, to be re-created in the big bang.

If the matter that collapses to form a black hole has a net electric charge, the resulting black hole will carry the same charge. This means that the black hole will tend to attract those members of the virtual particle-antiparticle pairs that have the opposite charge and repel those that have a like charge. The black hole will therefore preferentially emit particles with charge of the same sign as itself and so will rapidly lose its charge. Similarly, if the collapsing matter has a net angular momentum, the resulting black hole will be rotating and will preferentially emit particles that carry away its angular momentum. The reason a black hole "remembers" the electric charge, angular momentum and mass of the matter that collapsed and "forgets" everything else is that these three quantities are coupled to long-range fields: in the case of charge the electromagnetic field and in the case of angular momentum and mass the gravitational field.

Experiments by Robert H. Dicke of Princeton University and Vladimir Braginsky of Moscow State University have indicated that there is no long-range field associated with the quantum property designated baryon number. (Baryons are the class of particles including the proton and the neutron.) Hence a black hole formed out of the collapse of a collection of baryons would forget its baryon number and radiate equal quantities of baryons and antibaryons. Therefore when the black hole disappeared, it would violate one of the most cherished laws of particle physics, the law of baryon conservation.

Although Bekenstein's hypothesis that black holes have a finite entropy requires for its consistency that black holes should radiate thermally, at first it seems a complete miracle that the detailed quantum-mechanical calculations of particle creation should give rise to emission with a thermal spectrum. The explanation is that the emitted particles tunnel out of the black hole from a region of which an external observer has no knowledge other than its mass, angular momentum and electric charge. This means that all combinations or configurations of emitted particles that have the same energy, angular momentum and electric charge are equally probable. Indeed, it is possible that the black hole could emit a television set or the works of Proust in 10 leather-bound volumes, but the number of configurations of particles that correspond to these exotic possibilities is vanishingly small. By far the largest number of configurations correspond to emission with a spectrum that is nearly thermal.

The emission from black holes has an added degree of uncertainty, or unpredictability, over and above that normally associated with quantum mechanics. In classical mechanics one can predict the results of measuring both the position and the velocity of a particle. In quantum mechanics the uncertainty principle says that only one of these measurements can be predicted; the observer can predict the result of measuring either the position or the velocity but not both. Alternatively he can predict the result of measuring one combination of position and velocity. Thus the observer's ability to make definite predictions is in effect cut in half. With black holes the situation is even worse. Since the particles emitted by a black hole come from a region of which the observer has very limited knowledge, he cannot definitely predict the position or the velocity of a particle or any combination of the two; all he can predict is the probabilities that certain particles will be emitted. It therefore seems that Einstein was doubly wrong when he said, "God does not play dice." Consideration of particle emission from black holes would seem to suggest that God not only plays dice but also sometimes throws them where they cannot be seen.

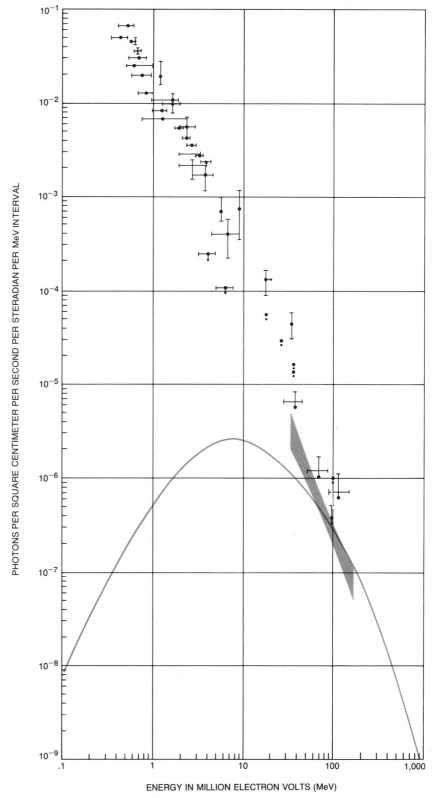

PRIMORDIAL BLACK HOLES, each about the size of an elementary particle and weighing about a billion tons, may have been formed in large numbers shortly after the big bang, the beginning of the universe as we know it. Such black holes would have a temperature of about 70 billion degrees Kelvin, corresponding to an energy of 10 million electron volts (MeV). The particles emitted at that energy would produce a diffuse spectrum of gamma rays detectable by satellites. The data points and the shaded region represent actual measurements of the diffuse gamma-ray spectrum in nearby space. The measurements indicate that the average density of such black holes in the universe must be less than about a million per cubic light-year. Solid curve is predicted spectrum from such a density of primordial black holes, based on plausible assumptions about the density of matter in universe and distribution of black holes.

# 10

# Will the Universe
# Expand Forever?

by J. Richard Gott III, James E. Gunn, David N. Schramm
and Beatrice M. Tinsley
*March 1976*

*The recession of distant galaxies, the average density
of matter, the age of the chemical elements and the
abundance of deuterium together suggest that the
expansion cannot be halted or reversed*

Cosmological inquiry is ancient, but only in the past 50 years or so have we begun to understand how the universe began and what its ultimate fate may be. The crucial perception came in the 1920's, when Edwin P. Hubble demonstrated that the spiral nebulas are not local objects but independent systems of stars much like our own, and thereby showed that the universe is a much larger place than had been imagined. Hubble showed further that the entire observable system of galaxies is in orderly motion. As is now well known, the nature of that motion is expansion: all distant galaxies are receding from us.

That the universe is expanding is today considered established. A question that remains unsettled is whether the expansion will continue forever or whether the receding galaxies will someday stop and then reverse their motion, eventually falling together in a great collapse. The answer to this question determines the geometrical character of the universe, that is, it determines the nature of space and time. If the expansion continues perpetually, the universe is "open" and infinite; if it will someday stop and reverse direction, the universe is "closed" and of finite extent.

In order to choose between those possibilities, astronomers construct mathematical models of the universe and then attempt to find observable features of the real universe that would confirm one of the models and exclude all others. So far no single measurement has been made with enough precision to settle the question unambiguously. Several independent tests are possible, however, and pieces of the puzzle have been supplied by many workers employing quite different techniques. It now seems feasible to assemble the pieces. Taken together, the available evidence suggests that the universe is open and that its expansion will never cease.

## Isotropic Expansion

Hubble detected the recessional motion of the distant galaxies through measurements of their optical spectra. The spectra of most stars (and hence of galaxies) are interrupted by dark lines representing the absorption of particular wavelengths by atoms in the cooler, outer layers of the stellar atmosphere; each chemical element generates a characteristic pattern of lines whose wavelengths are precisely known from laboratory measurements. When the galaxy is moving away from the observer, the wavelength of each spectral line is increased as a result of the Doppler effect, so that all the lines appear to be displaced toward longer wavelengths and in particular toward the red end of the visible portion of the spectrum. The displacement is called a red shift, and by measuring its magnitude the velocity of recession can be calculated. When an object is moving toward the observer, the wavelengths of the spectral lines are decreased by the Doppler effect and the lines appear to be displaced toward the blue end of the spectrum, an effect called a blue shift. All the distant galaxies whose spectra were measured by Hubble and by later observers show red shifts; they are therefore assumed to be receding from us.

The recessional motion has several remarkable properties. Hubble showed that the velocity with which a galaxy recedes is proportional to its distance from us, so that a constant ratio of distance to velocity can be calculated. The ratio is such that a galaxy 10 million light-years from us recedes with a velocity of 170 kilometers per second; another galaxy twice as far away recedes twice

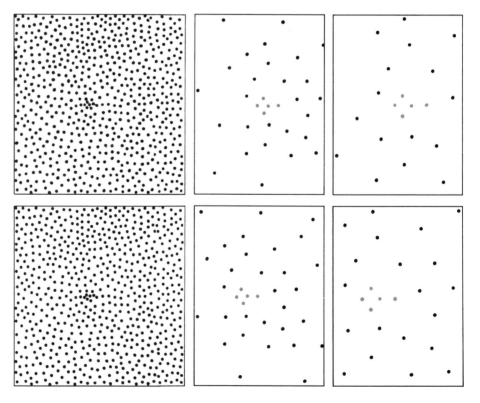

REMOTE PAST ————————————————————————————→ PRESENT ————

**FATE OF THE UNIVERSE is described through mathematical models of its behavior. Two classes of models are generally considered plausible; in both the universe begins in a compact state of infinite density (the big bang). In one class of models the universe expands continuously and indefinitely, albeit at an ever lower rate (*upper series of drawings*). In the other class**

as fast, or 340 kilometers per second [*see illustrations on next two pages*]. Small departures from this rule are commonly observed because most galaxies are members of groups or clusters and have orbital motions with a component along the line of sight connecting the earth with the galaxy. Those motions are essentially random, however, so that in any large sample of galaxies they cancel one another. Nonrandom, systematic variations from the ratio have been found only for galaxies at the most extreme distances; as we shall see, these variations do not invalidate the rule but provide important information about the history of the universe.

A second characteristic of the cosmic expansion is its isotropy: it is the same in all directions. No matter where in the sky a galaxy is found, its recessional velocity is related to its distance by the same proportionality. This observation seems to suggest that the universe is remarkably symmetrical and, what is even more extraordinary, that we happen to be at its very center. The crystal spheres of medieval cosmologies were no more geocentric.

There is, of course, another explanation, which can be understood most readily by considering a simple two-dimensional model of an expanding universe. Imagine a spherical balloon with small dots painted on its surface, each dot representing a galaxy. As the balloon is inflated the distance between any two dots (always measured on the surface of the sphere) increases with a speed proportional to the distance between them. No matter which dot is designated the center, all the other dots recede from it uniformly in all directions. Thus each dot observes the same expansion and no one of them has a privileged position. Such an expansion has no center; more precisely, every point is its center.

It follows from this analysis of the expansion that the geometrical configuration of the dots cannot change. A balloon bearing a picture of Mickey Mouse continues to bear the same picture as it is inflated. All distances between points on the balloon are multiplied by the same factor. Similarly, in the real universe eight galaxies that happen to lie at the corners of a cube in one epoch will remain at the corners of a cube, albeit a larger one, as the universe expands.

## The Big Bang

Since Hubble's original discovery increasingly precise observations have shown that it is not only the cosmic expansion that is isotropic; all the large-scale features of the universe are indifferent to direction. For example, the distribution of galaxies on the celestial sphere and the distribution of extragalactic radio sources appear to be quite uniform. The most compelling evidence of isotropy was discovered in 1965 by Arno A. Penzias and Robert W. Wilson of the Bell Laboratories; it is the microwave background radiation that seems to bathe the entire universe. The microwave radiation has since been shown to be highly isotropic; it varies by less than one part in 1,000 over the entire area of the sky.

The observation of such remarkable isotropy has led to the adoption of a powerful generalization called the cosmological principle. It states that the universe appears isotropic around all observers participating in the expansion everywhere and at all times. In other words, our galaxy is indeed at the center of the universe, but so is every other galaxy.

The cosmological principle also governs the behavior of the two-dimensional model universe represented by a spherical balloon. If the painted dots are distributed with uniform density over the surface of the balloon, the neighborhood of any chosen point is statistically the same as that of any other point and no direction has any special significance. Indeed, it is not necessary to postulate independently that the dots (or, in the three-dimensional universe, the galaxies) are uniformly distributed. If the universe is isotropic for all observers, then the distribution must be homogeneous; if it were not, an observer at the edge of a density fluctuation would not see a uniform distribution independent of direction.

For the purposes of this discussion we shall adopt the cosmological principle, but it must be remembered that its appeal is mostly philosophical. It has not been ade-

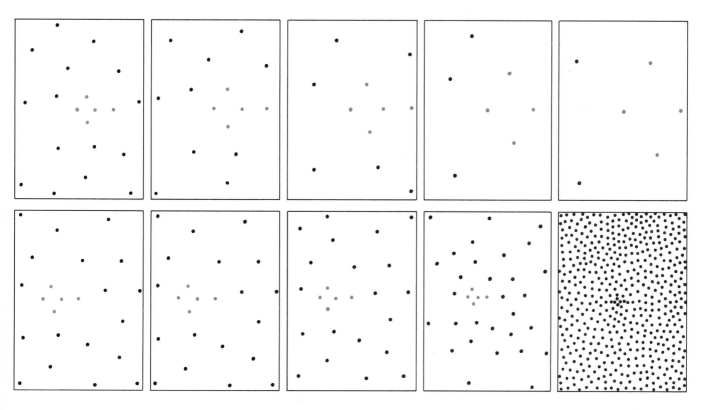

→ REMOTE FUTURE

the universe expands to a maximum size, then begins to contract, eventually reaching infinite density again (*lower series of drawings*). The alternatives are illustrated here in an arbitrary region of space in which the cosmic expansion is represented by a decrease in average density. The expansion is isotropic, that is, the same in all directions. As a result an observer at any point perceives himself as being at the center of the expansion, and the shape of any pattern (such as the arbitrary pattern of colored dots) will be preserved in all epochs.

quately tested, and indeed adequate tests may never be possible.

Given our knowledge of the universe as we observe it today, what can we deduce about its history? A simple, hypothetical model with which to begin is one where the recessional velocity of every galaxy has remained unchanged through all time. It is then apparent that any galaxy now receding from us was once arbitrarily close, and that the time that has elapsed since then is equal to the ratio between the galaxy's distance and its velocity. Since the ratio is the same for all galaxies, all of them must have been nearby at the same time; in other words, at some unique time in the past all the matter in the universe was compressed to an arbitrarily great density. The time calculated to have elapsed since that compact state existed, assuming the rate of expansion has not changed, is called the Hubble time. Its reciprocal, by which one multiplies the distance of a galaxy to obtain its recessional velocity, is called the Hubble constant. Measurements of the Hubble time are complicated by uncertainties as to the distance to galaxies, and the measurements have been repeatedly revised since Hubble's first estimate of about two billion years. The Hubble time is now thought to be between 12 and 25 billion years, and the most likely value is about 19 billion years.

If the motions of the galaxies are extrapolated into the past as far as possible, a state is eventually reached in which all the galaxies were crushed together at infinite density. That state represents the big bang, and it marks the origin of the universe and everything in it. By the simple mathematics of

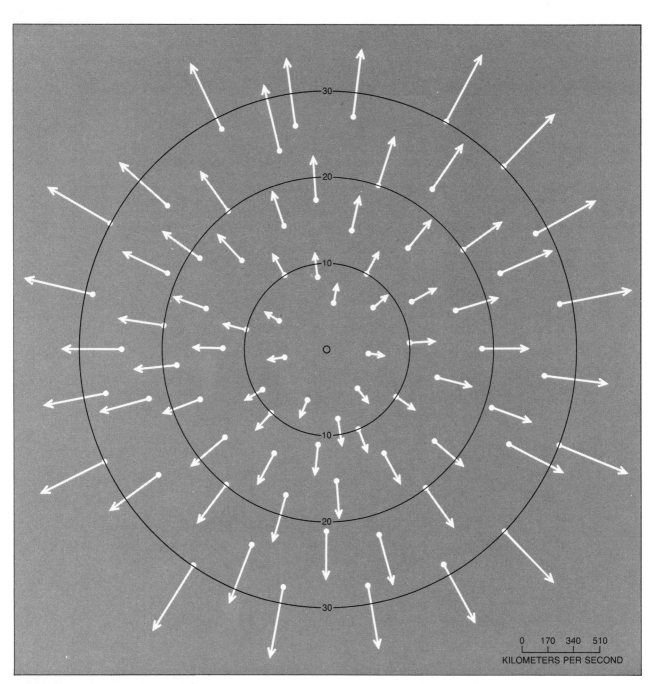

COSMIC EXPANSION seems to place the observer at the center of the universe, from which all distant galaxies are fleeing. The velocity with which a galaxy recedes is proportional to its distance from the observer, a relation first established in the 1920's by Edwin P. Hubble from observations made with the 100-inch telescope at the Mount Wilson Observatory. The principle that the ratio of velocity to distance is constant has since become known as Hubble's law. It can be interpreted most simply as evidence that the expansion of the universe began with the big bang, since the relation implies that in the past all the galaxies were packed together with infinite density. Distances are given here in millions of light-years; velocities are represented by the lengths of arrows, measured on the scale at lower right.

proportionality, if the recessional velocities have not changed, the date of the big bang must be exactly one Hubble time ago. Actually the rate of expansion has almost certainly not been constant, but that does not alter the fact of the big bang; it merely changes the date.

That the universe began with a big bang is an inevitable conclusion if the known laws of physics are assumed to be correct and in some sense complete. It is conceivable, however, that there are laws of nature whose effects are negligible on the scale of the physics laboratory, or even on the scale of the solar system, but that might predominate in determining the behavior of the universe as a whole. One theory requiring new laws of nature was the steady-state cosmology, in which the universe is unchanging and infinitely old. In order to explain the cosmic expansion the steady-state theory postulated the continual creation of matter from the void.

In the steady-state model of the universe it was particularly difficult to account for the microwave background radiation. This radiation field has the spectral characteristics of the thermal radiation emitted by a body at a temperature of 2.7 degrees Kelvin. It seems to be satisfactorily explained only as a relic of an epoch when the universe was very hot and very dense. A steady-state universe cannot have had such conditions, since in that model all conditions, by definition, have not changed.

In big-bang models the background radiation is a natural consequence of conditions in the early universe. The initial state in these models is one of high temperature and density, a state sometimes called the cosmic fireball. At this stage the matter and the electromagnetic energy composing the universe are thought to have been in thermodynamic equilibrium, and as a result the radiation spectrum was that of a very hot body. As the universe has expanded the radiation has cooled, reaching the low-temperature spectrum observed today. The cooling can be understood as an enormous red shift; since all galaxies are constantly receding from the radiation, its spectrum is constantly displaced toward the longer wavelengths associated with lower energies and lower temperatures. In 1946 George Gamow predicted the existence of a thermal background radiation entirely from the theoretical framework of the big-bang model. He estimated its present temperature as being about five degrees K. The general agreement between Gamow's prediction and the observations of Penzias and Wilson is the most compelling evidence for the big bang.

Thus it appears that the universe began from a state of infinite density about one Hubble time ago. Space and time were created in that event and so was all the matter in the universe. It is not meaningful to ask what happened before the big bang; it is somewhat like asking what is north of the North Pole. Similarly, it is not sensible to ask where the big bang took place. The point-universe was not an object isolated in space; it was the entire universe, and so the

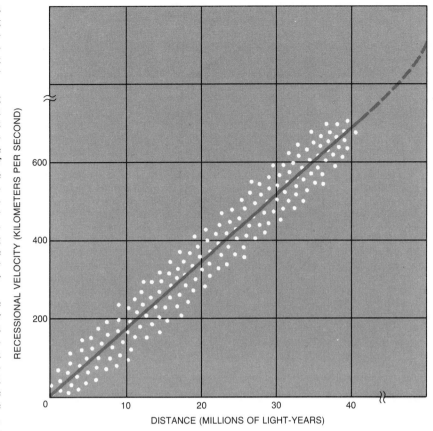

HUBBLE'S LAW is established by measuring the ratio of velocity to distance for many galaxies. The best estimate of the ratio (solid colored line) is about 17 kilometers per second per million light-years. Individual galaxies (white dots) depart from that value because most are members of clusters and have orbital velocities. The inverse of the ratio is the Hubble time: the time it would have taken for any given galaxy at its present velocity to reach its present position, or in other words the time since the big bang if the velocities have not changed. Actually it is thought that the recessional velocities have declined under the influence of gravitation; as a result the ratio is thought to increase at extreme distances (broken colored line).

only answer can be that the big bang happened everywhere.

In most models of the evolving universe the receding galaxies are assumed to follow ballistic trajectories, roughly analogous to those of a thrown ball or an artillery shell. The galaxies were propelled apart by forces acting at the moment of the big bang, but since then they have moved in free flight, without further propulsion. They should therefore continue in uniform motion if no other forces acted on them. Actually the galaxies continue to interact as they fly apart. If only the familiar forces that express the known laws of physics are to be allowed in our models, then only one force can have a significant effect on the expansion: gravitation. We can therefore hope to understand the dynamics of an expanding universe if we can describe the gravitational interactions of all its components.

## Gravitational Deceleration

The gravitational force affects all matter, it is always attractive and its range is infinite. Moreover, gravitation has a peculiar geometrical property that significantly aids analysis: A hollow sphere exerts no net gravitational forces on masses in its interior. (Actually, of course, the mass of the hollow shell attracts the masses in the interior, but all the forces exactly cancel, so that at every point in the interior the resultant force is zero.) This proposition was first proved by Newton, but it applies equally well to more recent theories of gravitation, such as the general theory of relativity.

If a spherical region of the universe is selected for examination, the rest of the universe surrounding it can be regarded as a hollow spherical shell, since the cosmological principle requires that the surrounding matter be uniformly distributed in all directions. The selected sphere can then be studied as if it were isolated and not subject to forces from the outside. The cosmological principle also ensures that any selected sphere of galaxies will expand or contract by the same factor as the universe as a whole, regardless of the sphere's location or size. In order to characterize the dynamics of the universe, therefore, we need only examine the dynamics of a representative sphere within it. If the sphere chosen is a small one, the velocities of the galaxies will be much smaller than the speed of light, and their motions can be described in terms of Newtonian mechanics.

A galaxy at the edge of such a small

spherical region feels only the gravitational forces generated by the matter inside the sphere. If that matter is distributed homogeneously, then the resultant force acting on the galaxy attracts it to the center of the sphere. As a consequence the test galaxy does not recede with constant velocity; instead its recessional motion is at all times decelerated. It is thus obvious that in the past the test galaxy and all other galaxies must have been moving faster than they are today. Ignoring the deceleration leads to an overestimate of the age of the universe. That age is one Hubble time only if the rate of expansion has not changed; since the rate has slowed under the influence of gravitation, the big bang must have taken place more recently than one Hubble time ago.

The magnitude of the gravitational deceleration clearly depends on the amount of mass inside the selected sphere. If the sphere contains a great deal of matter, the test galaxy must eventually stop and fall toward the center; the representative spherical region begins to collapse and, on the cosmological principle, so does the entire universe. If there is little matter, the test

galaxy will decelerate continuously but never stop. Both the spherical region and the universe as a whole will expand indefinitely. The situation is analogous to that of a projectile shot upward from the surface of the earth: the projectile decelerates but nevertheless will not return to the surface if its velocity exceeds a certain critical value, the escape velocity.

The escape velocity for objects leaving the earth is determined by the mass of the earth; for a test galaxy at the edge of an arbitrary sphere in space the escape velocity is determined by the total mass within the sphere. From the ratio of velocity to distance the actual velocity of the test galaxy with respect to the center of the sphere is known. Its ultimate fate therefore depends on the value of the escape velocity and hence on the mass within the sphere.

Since the universe is assumed to be homogeneous, the determining quantity is the average density of matter in the universe. If the density is smaller than some critical value, the effect of gravitation is too small to halt the cosmic expansion, and all galaxies will recede forever (although ever more

slowly). If the density is greater than the critical density, gravitation will prevail, and the expansion will slow to a stop and begin an accelerating contraction ending in a final cataclysm: what might be called the big crunch. The actual value of the critical density depends on the Hubble time, which is not precisely known. If the Hubble time is 19 billion years, the critical density is $5 \times 10^{-30}$ gram per cubic centimeter, the equivalent of about three hydrogen atoms per cubic meter. That seems to be an exceedingly small density, but it should be remembered that on the average the universe is quite empty.

The effect of gravitation on the cosmic expansion can be incorporated into mathematical models most conveniently by introducing a dimensionless number called the density parameter and denoted by the Greek letter omega ($\Omega$). The density parameter is defined as the ratio of the actual density of the universe to the critical density. If the universe is to expand forever, that ratio must be less than or equal to 1; if $\Omega$ equals exactly 1, the universe is expanding everywhere at just the escape velocity, and if $\Omega$ is greater than 1, the universe must eventually collapse.

## The Geometry of Space

The foregoing discussion could have been derived entirely from the Newtonian theory of gravitation, although it is also valid in the general theory of relativity. In the general theory, however, the value of the density parameter has further consequences; in particular it determines the geometry of space. In the high-density universe fated to collapse, gravitation is sufficiently strong to "close" space. The total volume of the universe is finite at all times, although there is nevertheless no boundary or edge to the universe. A two-dimensional analogue of such a three-dimensional space is the surface of a sphere, which similarly is finite in area although it has no boundary.

If $\Omega$ equals 1, so that the universe expands with exactly the escape velocity, the geometry of space is "flat"; it is the familiar Euclidean geometry, and it is represented in two dimensions by an infinite plane.

The geometry of a perpetually expanding universe in which $\Omega$ is less than 1 is more difficult to illustrate. The two-dimensional analogue is the surface of a figure called a pseudosphere, and a complete pseudosphere cannot be constructed in three-dimensional space. A saddle-shaped surface has some of the properties of such a space, but it is a defective model in the important respect that it has a center, whereas the real space defines no preferred position [see illustration on page 88]. Perhaps the best two-dimensional representation of such a space is a projection of a pseudosphere onto a plane, a device that is employed in several of the works of the Dutch artist M. C. Escher [see illustration on page 89].

The three possible kinds of three-dimensional space are distinguished by several ge-

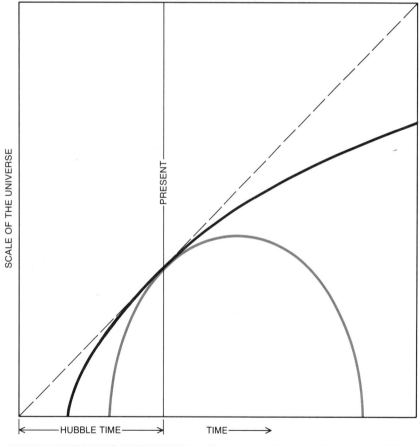

MODELS OF COSMIC EVOLUTION describe changes in the scale of the universe with the passage of time. All models must be consistent with the scale and the rate of expansion observed today, so that all their graphs must be tangent at the present moment. If the rate of expansion is unchanging (*broken black line*), the age of the universe is the Hubble time. Decelerating universes are younger, and both their history and their destiny depend on the magnitude of the deceleration. With modest deceleration the expansion can continue indefinitely, albeit at an ever lower rate (*solid black line*). Greater deceleration implies that the cosmic expansion must stop and then reverse, leading to an eventual collapse (*colored line*). The infinitely expanding universe is said to be "open" and the collapsing universe, which is also the youngest, "closed."

| | | OPEN | CRITICAL | CLOSED |
|---|---|---|---|---|
| DENSITY PARAMETER $\Omega$ | $\dfrac{\text{ACTUAL DENSITY}}{\text{CRITICAL DENSITY}}$ | $\Omega < 1$ | $\Omega = 1$ | $\Omega > 1$ |
| DECELERATION PARAMETER $q_0$  DECELERATION | $\dfrac{\text{DISTANCE}}{(\text{VELOCITY})^2}$ | $q_0 < \frac{1}{2}$ | $q_0 = \frac{1}{2}$ | $q_0 > \frac{1}{2}$ |
| GEOMETRY OF SPACE | | HYPERBOLIC (NEGATIVE CURVATURE) | FLAT (ZERO CURVATURE) | SPHERICAL (POSITIVE CURVATURE) |
| FUTURE OF THE UNIVERSE | | PERPETUAL EXPANSION | PERPETUAL EXPANSION | EVENTUAL COLLAPSE |

OPEN AND CLOSED MODELS of the universe are distinguished mainly by the average density of matter and by the value of the cosmic deceleration. Density is a crucial factor because in models described by the general theory of relativity it is the sole determinant of the gravitational forces that slow the cosmic expansion. Density is most easily treated as a dimensionless parameter, the ratio of the actual density to the critical density just needed to halt the expansion. Deceleration can also be expressed as a dimensionless number, the deceleration parameter, which in the models considered here is always equal to half the density parameter. These two parameters determine not only the future of the universe but also the geometry of space. The open universe is of infinite size at all times, and in it space has hyperbolic, or negative, curvature. In the universe with critical density, in which the density parameter is exactly 1, space has zero curvature, it is the flat space of Euclidean geometry. The closed universe is of finite size; in it space has spherical, or positive, curvature.

ometric properties, some of which can be represented in the two-dimensional models. A flat plane, of course, is the basis of Euclid's geometry, and on it all the Euclidean axioms and the theorems derived from them are obeyed. On a plane exactly one line can be drawn through a given point parallel to another line; the sum of the included angles in a triangle is always 180 degrees; the circumference of a circle increases in proportion to the radius, and the area of a circle increases in proportion to the square of the radius.

On the surface of a sphere no two lines are parallel, provided that a straight line is defined as one taking the shortest path between two points. Such lines are called geodesics, and on the sphere they are the great circles, any two of which always intersect. Similarly, on a sphere the sum of the included angles in a triangle is always greater than 180 degrees; the circumference of a circle increases more slowly than in proportion to the radius, and the area of a circle increases more slowly than in proportion to the square of the radius.

The surface of a pseudosphere possesses properties opposite to those of a sphere. Through a given point infinitely many lines can be drawn that are parallel to another line, or geodesic. The sum of the angles of a triangle is less than 180 degrees. The circumference of a circle increases faster than in proportion to the radius, and the area of a circle increases faster than in proportion to the square of the radius. The geometry of the three-dimensional space represented by a pseudosphere was first studied in 1826 by Nikolai Lobachevski.

In the simple cosmological models discussed here the geometry of space is uniquely related to the future behavior of the universe. It is notable that in models with $\Omega$ greater than 1 the universe is closed in both space and time. The volume of space is finite, and there are definite temporal limits, beginning with the big bang and ending with the big crunch. Models in which $\Omega$ is less than or equal to 1 are open in both space and time. Such models have a definite starting point (the big bang), but they are always infinite in extent and they expand indefinitely into the future.

## Measurements of Deceleration

There are several possible ways of determining whether the actual universe is open or closed. All of them lead ultimately to an estimate of the rate at which the cosmic expansion is decelerating. One method is simply to measure the deceleration directly, by observing distant galaxies. It is also possible to measure the age of the universe, and by comparing it with the Hubble time (the age if there were no deceleration) to derive an estimate of how much the velocity of expansion has changed. Since the deceleration is a gravitational phenomenon, an equivalent measure is the average density of matter; comparing the actual density with the critical density gives the ratio $\Omega$. Finally, the present abundance of certain chemical elements represents a kind of fossil record of conditions in the very early universe, including the density, and from that information too the value of $\Omega$ can be calculated. Evidence from each of these methods has contributed to our present knowledge of the state of the universe.

The deceleration of the cosmic expansion is usually expressed in terms of a dimensionless number called the deceleration parameter and symbolized $q_0$. Since the deceleration is a gravitational effect, the deceleration parameter is closely related to the average density of matter. In the cosmological models considered here, which are constructed according to the general theory of relativity, $q_0$ is always equal to exactly half the density parameter $\Omega$. Thus if $q_0$ is greater than 1/2, the universe is decelerating rapidly enough, because of its high den-

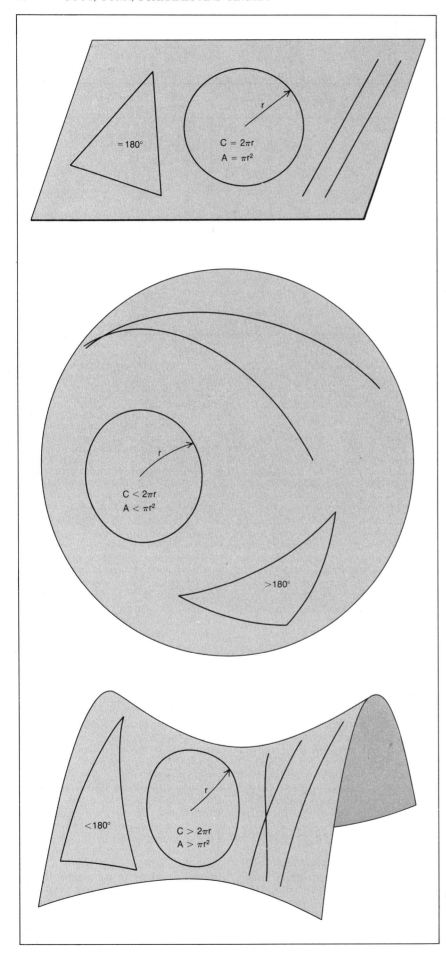

sity, to stop expanding and subsequently collapse. If $q_0$ is less than 1/2, the expansion cannot stop because the density is too low to halt it.

An obvious approach to determining the rate of deceleration would be to measure the radial velocity of a single galaxy at two times in order to see how much it has slowed down in the interval. Unfortunately the change in velocity expected in a human lifetime is far too small to be detected; indeed, the experimental errors involved in the determination are many orders of magnitude larger. Because of the finite speed of light, however, it is possible to measure the velocities of galaxies in the remote past and to compare them with velocities representing more recent eras. The comparison is possible because as we look at increasingly distant objects in the sky we also see farther and farther into the past. The relation is obvious when distances are measured in light-years: if a galaxy is a billion light-years away, the light we receive from it today was emitted a billion years ago, and the Doppler shift in its spectrum must reflect the distant galaxy's velocity then with respect to our own velocity now. Thus if the cosmic expansion is decelerating, the constant ratio of velocity to distance discovered by Hubble is not expected to hold for the most distant galaxies. At extreme distances the ratio should increase, or in other words the velocities should be greater than those predicted by Hubble's law.

In order to measure the deceleration by this method it is necessary to have an independent measure of the distances of the galaxies. For all but the nearest galaxies the only practical method of estimating the distance to a galaxy is from its apparent luminosity. If all galaxies at all times had the same intrinsic luminosity, then their apparent brightness would vary simply as the inverse of the square of their distance, and the calculation of distance would be a straightforward procedure. Of course, they do not all have the same intrinsic luminosity.

Random variations in brightness (caused, for example, by differences in size) may produce errors in any individual measurement. Because of such variations it is necessary to acquire a large volume of data and to submit it to statistical analysis, but in principle random variations are not a serious con-

**GEOMETRY OF SPACE** characteristic of each model universe has an analogous surface. The properties of the surfaces are defined by the Euclidean axioms and theorems on parallel lines, on the included angle of a triangle and on the circumference and area of a circle. The flat space of a critical universe is represented by a plane, and the positively curved space of a closed universe corresponds to the surface of a sphere. Some of the properties of the negatively curved space of an open universe can be demonstrated on a saddle-shaped surface, but the saddle is an imperfect analogue because it has a center. The best representation of an open universe is an infinite surface called a pseudosphere, which cannot be constructed in a three-dimensional space.

SURFACE OF A PSEUDOSPHERE is represented in an etching, *Circle Limit IV*, by M. C. Escher. In the etching the surface is projected onto a plane. As in any map projection, the scale is not constant; on the pseudosphere itself the figures of angels and demons would all be the same size. If a single figure is regarded as a unit of measure, it is apparent that the circumference of a circle increases much faster than in proportion to the radius. Similarly, each figure defines a triangle (with the vertexes at the feet and the wing tips); from the number of triangles that meet at each vertex it can be shown that on the pseudosphere the sum of the angles of a triangle is less than 180 degrees. The pseudosphere is an infinite surface of negative curvature, analogous to space in a universe that expands forever. It has no privileged position that could be considered a center, and projection would be unchanged if it were centered on any other point.

cern, since in any large sample they can be expected to cancel out. Systematic variations, however, require explicit correction.

Theories of stellar evolution suggest that the combined light from all the stars in an isolated galaxy probably declines at a rate of a few percent per billion years. Galaxies were therefore probably brighter in the remote past. If the change in brightness were neglected in making measurements of the deceleration, the calculated distances would be too small and as a result the rate of deceleration would be overestimated. The decline in brightness would seem to be quite modest, but it changes the calculated value of the deceleration parameter $q_0$ by about 1, which is more than enough to decide between an open universe and a closed one. The best current observations, which take into consideration the changes in luminosity resulting from stellar evolution, suggest that $q_0$ is closer to zero than to 1/2 and therefore that the universe is open and perpetually expanding.

There is a further large uncertainty in the determination of the deceleration. Most of the observed galaxies are situated in relatively dense clusters, and possible interactions between galaxies ought to be taken into account. For example, it has recently been shown that in clusters large galaxies swallow smaller ones, with a consequent change in luminosity and size. It is not yet possible to predict the magnitude of the change, or even to be sure whether it makes the measured luminosity increase or decrease. Adding stars to a galaxy should make it brighter, but in cosmological observations only the luminosity of the central part of the galaxy is measured. If the cannibal galaxy swells significantly, the number of stars in the central region might be reduced and the galaxy would appear fainter.

## The Age of the Universe

As a result of statistical uncertainties and our imperfect knowledge of galactic evolution the value of $q_0$ derived from measurements of recessional velocity is very uncertain. From this test alone one cannot conclude that $q_0$ is less than 1/2 and that

the universe is open; on the other hand, very large values of $q_0$, such as $q_0$ equals 2, do seem to be excluded.

The second approach to determining the fate of the universe is to measure its age. If the expansion were not decelerating at all, the age would be equal to the Hubble time. Since it is decelerating it must be somewhat younger than the Hubble time. By finding the difference between the actual age and the Hubble time it is possible in principle to calculate the deceleration parameter $q_0$.

The age of the universe can be estimated by two methods; both actually yield only lower limits, since they measure the ages of objects in the universe, but those objects were probably formed within the first billion years or so after the big bang. The first method consists in determining the age of the oldest stars that can still be observed today. The oldest stars close enough for detailed observation are thought to be those in the globular clusters associated with our own galaxy. Models of stellar evolution indicate that they are between eight and 16 billion years old.

The age can also be estimated from measurements of the relative abundance of certain heavy elements. All the elements heavier than iron, including several radioactive ones, are thought to have been formed in supernovas, which have probably been exploding in the galaxy since its formation. Because each radioactive element decays at a constant rate, the ratio of the abundance of each radioactive element to the abundance of its decay products can reveal the average age of the heavy elements. The ratios indicate that the age of the galaxy is between six and 20 billion years. The two calculated ages are thus consistent, and they suggest that the big bang took place between eight and 18 billion years ago.

## Average Density

Whether a given age within the allowed range corresponds to an open universe or a closed one depends on the value of the Hubble time, and as we have seen that value is not easily determined. Moreover, even if the Hubble time is assumed to equal the recent best estimate of 19 billion years, neither the age limits nor the exclusion of $q_0$ values greater than 2 is sufficient to decide whether the universe is open or closed [see illustration on page 91]. The issue can be decided only by imposing further constraints.

The third test consists in measuring the average density of matter in the universe and thereby deriving the density parameter $\Omega$. A lower limit to the density can be obtained by considering only the mass associated with visible galaxies. The density is found by counting the galaxies in a given volume of space, multiplying by the masses of the galaxies and dividing by the volume.

Weighing a galaxy is not as difficult as it might at first seem. Few galaxies are completely isolated; most are found in small groups or in large clusters, and their mass can be deduced from observations of the gravitational effects they exert on one another. Two galaxies in orbit around each other, for example, must have a gravitational attraction just sufficient to balance the centrifugal force. If their separation and their velocities with respect to each other are known, the determination of their combined mass becomes a simple exercise in Newtonian mechanics. The procedure for clusters of many galaxies is only slightly more complicated. Significantly, the mass calculated in this way includes not only the mass of the galaxies but also the mass of any other matter in the cluster. Constituents that would not be visible, such as black holes or extragalactic dust and gas, are automatically taken into account.

Estimates of the mass of a great many galaxies, combined with counts of the galaxies in large volumes of space, give an indication of the value of the density parameter $\Omega$. If the mass associated with galaxies represents all the mass in the universe, then $\Omega$ is only about .04 and the universe must definitely be open and infinitely expanding. This value could be uncertain by a factor of 3, so that a value of $\Omega$ as great as .12 would still be consistent with observations, but that is still well below the value of 1 required to close the universe.

The density of the universe can also be estimated by comparing the behavior of distant galaxies with the behavior of those in the local supercluster, the aggregate of galaxies that includes our own local group in addition to many other small groups and the somewhat larger Virgo cluster. Within the local supercluster the mean density of galaxies is two and a half times greater than that in the universe as a whole. If all mass is associated with galaxies, then the average density of matter must also be two and a half times greater in the supercluster than outside it. The difference in density should give rise to a difference in the rate of expansion; because the local density is greater, nearby galaxies should be more strongly decelerated. The magnitude of the difference depends on the value of $\Omega$; if $\Omega$ is large, there should be a considerable difference. If $\Omega$ is small, then the deceleration everywhere is small, and even a local enhancement in density by a factor of two and a half would cause little change. In fact, the difference is undetectable, being smaller than the probable observational errors. The most straightforward conclusion is that $\Omega$ is very small, probably no larger than .1.

Both methods of estimating density are explicitly confined to the matter associated with galaxies, and an obvious objection to them is that there might be substantial amounts of matter elsewhere in the universe. That possibility cannot be excluded, but there is no evidence to substantiate it.

Current theories show that clusters of galaxies could have formed from a universe in which matter was distributed much more smoothly than it is now. Debris left over from the formation of galaxies would also be collected by the clusters. Thus any particles that are not now in the clusters must have preferentially avoided them, that is, the particles must have had the special and unusual initial positions and velocities that would enable them to escape capture. Even if a large amount of matter were distributed uniformly outside the clusters now, it would fall into them in a few billion years.

Alternatively, the necessary mass could consist of some uniformly distributed medium with enough internal pressure to be unaffected by the gravitation of galaxies. It might, for example, be made up of large numbers of neutrinos or of gravitational waves. There is, however, a strong argument against such a pervasive "radiation-like" medium: it would almost certainly have prevented galaxies and clusters of galaxies from ever forming.

The density of all matter in the universe, whether or not it is associated with galaxies, can in principle be determined, but only by extrapolating from conditions in the present universe to those a few minutes after the big bang. The simplest assumptions about that early period suggest that the temperature and density were high enough for some subatomic particles to interact and form sizable amounts of some of the lighter nuclei. In particular a proton and a neutron could fuse to make a nucleus of deuterium, and most of the deuterium nuclei would quickly combine to form helium nuclei, composed of two protons and two neutrons. The proportion of deuterium and helium formed in this way depends on the density of the universe at the time when it was hot enough for the reactions to take place. From the early density and from the present temperature of the microwave background radiation it is possible to deduce the density today.

## Primeval Density

Mathematical models indicate that for the entire possible range of densities in the early universe between 20 and 30 percent of the matter is converted into helium. The helium abundance measured in a variety of astronomical objects is in this narrow range, which strongly supports the fundamental assumption that the universe went through a period of extreme temperature and density shortly after the big bang. The present abundance of deuterium depends strongly on the early density [see illustration, page 92]. The relative abundance of deuterium in nearby interstellar space has been measured by the third Orbiting Astronomical Observatory satellite, named Copernicus. After taking account of the deuterium depleted by nuclear reactions in stars, the measured abundance yields an average present density of about $4 \times 10^{-31}$ gram per cubic centimeter. The measure-

DECELERATION of the cosmic expansion can be detected in the recessional velocities of galaxies in the remote past. It is possible to look into the past by observing the most distant galaxies, since light reaching us now was emitted a length of time ago given by the galaxy's distance in light-years. The deceleration is thus perceived as a departure from Hubble's law; if there were no deceleration, the ratio of velocity to distance would be constant (*black line*); with deceleration the ratio increases at extreme distances (*colored line*). Because of the difficulty of estimating the distance to galaxies it has not been possible to measure the rate of deceleration precisely, but values of the deceleration parameter greater than about 2 have been excluded.

ment is a sensitive indicator of density: if the universe were 10 times denser, the big bang would have made less than a thousandth the observed abundance of deuterium. For this reason uncertainties in the measurement do not result in large uncertainties in the estimated density.

Whether the density determined by the deuterium abundance represents an open universe or a closed one depends on the Hubble time. As we have seen, if the Hubble time is 19 billion years, the critical density is $5 \times 10^{-30}$ gram per cubic centimeter, so that $\Omega$, the ratio of actual density to critical

density, is about .08. For any value of the Hubble time between 13 and 19 billion years, the value of $\Omega$ derived from the deuterium abundance is consistent with that derived from the density of galaxies. Conversely, for any plausible value of the Hubble time, a value of $\Omega$ as great as 1 is inconsistent with the density required for the manufacture of deuterium.

The abundance of deuterium would seem to provide powerful evidence that the universe is open; unfortunately the arguments supporting that conclusion are somewhat insecure. In extrapolating from the present

state of the universe to conditions soon after the big bang the simplest possible model has been employed; other models might allow the observed amounts of helium and deuterium to be made in a much denser, closed universe. Those models are more complicated, even somewhat contrived, but they cannot be excluded. Moreover, the significance of the deuterium abundance depends entirely on the assumption that all the deuterium in the universe was made shortly after the big bang. Other sources, such as supernovas, have been suggested, but so far no mechanism has been found that would

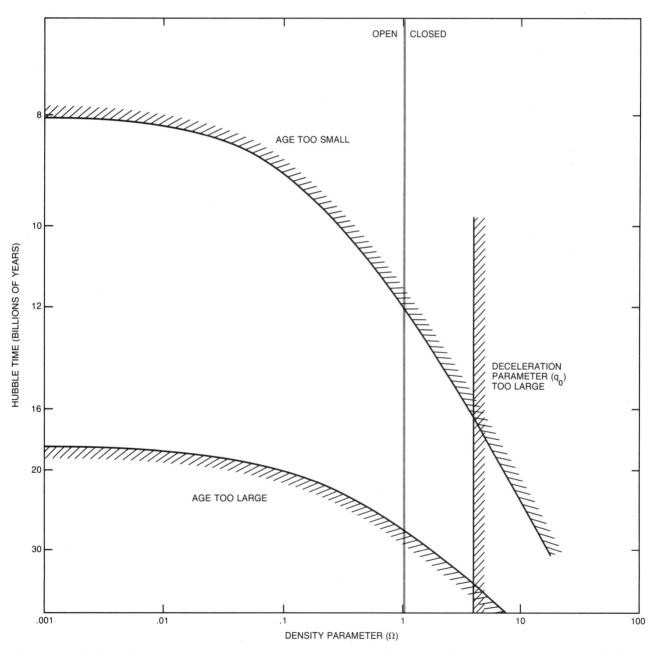

INITIAL CONSTRAINTS on the state of the universe are provided by determinations of its age and of the deceleration parameter. Estimates of the age of the oldest stars and of the average age of heavy elements suggest that the universe is between eight and 18 billion years old; the corresponding limits to the Hubble time depend on the density. Observations of distant galaxies provide an upper limit for the deceleration parameter: it cannot exceed 2, and the density parameter therefore cannot be greater than 4. The constraints derived from these measurements alone do not determine whether the universe is open or closed, since they encompass both kinds of model.

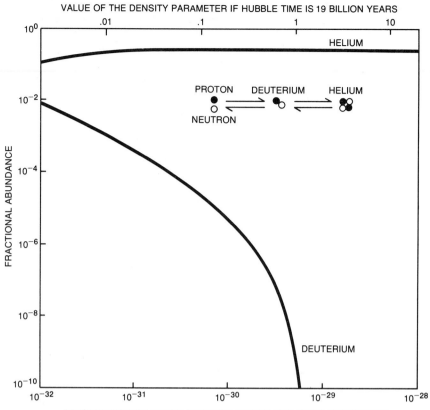

**DENSITY OF THE EARLY UNIVERSE** influenced the synthesis of deuterium and helium, and from the relative abundances of those elements the present density can be inferred. Deuterium is thought to have been formed by the fusion of protons and neutrons in the first few minutes after the big bang, but if the density then was too great, most or all of the deuterium would have been converted into helium. The abundances of both elements are shown as fractions (by mass) of all the matter in the universe. If the simplest models of the early universe are correct, and if deuterium has not been formed in more recent events, the observed abundance suggests that the density of the universe cannot be greater than about $4 \times 10^{-31}$ gram per cubic centimeter.

create a significant amount of deuterium without violating other constraints.

### Plausible Models

The measurements of the deceleration parameter, of the age of the universe, of the density of galaxies and of the abundance of deuterium all impose independent constraints on the state of the universe. If the measurements are consistent, there must be some class of models of the universe that is allowed by all the constraints. Indeed there is, and moreover it is a relatively small class, so that interesting predictions about the future of the universe are possible [*see illustration on page 93*]. If the universe is not too old, and if its density is at least equal to that observed in galaxies but not too great to make deuterium, the value of $\Omega$ must be between .04 and .09. That is far below the value required for a closed universe.

Two additional observations are consistent with the allowed values of $\Omega$ and the Hubble time. The calculated age of the stars in globular clusters is sensitive to the abundance of helium, and as we have seen that in

turn is determined by the density of the universe. It is therefore encouraging to find that the age and the helium abundance allowed by the combined constraints are compatible with what is known of the globular-cluster stars.

The constraints also require that the Hubble time itself be between 13 and 20 billion years. The direct determination of the Hubble time is difficult, but in recent years much effort has been expended on the problem by Allan R. Sandage and Gustav A. Tammann of the Hale Observatories. Their best value is $18 \pm 2$ billion years. Robert P. Kirshner and John Kwan of the California Institute of Technology have employed a different technique, relying on the properties of exploding stars in distant galaxies, to make an independent measurement of the Hubble time. They place the value between 13 and 22 billion years.

The consistency of the results obtained by such diverse methods is gratifying, and it encourages confidence that the cosmological model is well determined and the fate of the universe is known. Because of uncertainties in the data and in the theory em-

ployed to interpret them, however, the apparent agreement may yet prove to be fortuitous.

A firm prediction of the models considered here is that the deceleration parameter equals half the density parameter, and as we have seen this prediction cannot yet be tested. If in the future it is found to be wrong, more complicated cosmological models will be required. For example, one class of models employs a modification of general relativity once suggested by Einstein, in which a parameter called the cosmological constant is introduced. In these models space itself generates an attractive or repulsive gravitational force, and as a result the deceleration is no longer related in a simple way to the density.

Taken one at a time, each of the constraints we have discussed has possible loopholes. In particular, some of our colleagues would disagree with the small density derived from estimates of the mass associated with galaxies, and with the inclusion of a constraint on density derived from the production of deuterium. Our arguments and conclusions, however, derive their credibility from the fact that a consistent cosmological model can be constructed by the most straightforward interpretation of each piece of evidence. It is remarkable that such diverse factors as the age of stars, the mass of galaxies, the abundance of chemical elements and the observed rate of expansion of the universe can all be interpreted naturally in terms of one of the simplest cosmological models. This model describes a universe that is infinite in extent and that will expand forever. The case for an open universe is by no means closed, but it is strongly supported by the weight of the evidence.

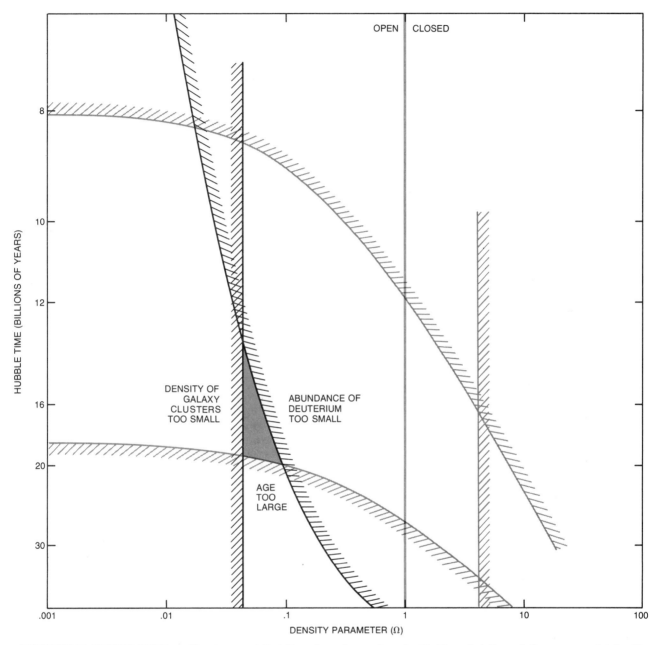

ADDITIONAL CONSTRAINTS combine to suggest that the universe will expand forever. The abundance of deuterium implies an upper limit to the density of all matter in the universe, and therefore also limits the density parameter, although the numerical value of that limit depends on the Hubble time. A maximum value of the Hubble time itself is defined by the estimates of the ages of stars and the heavy elements. Finally, calculations of the mass associated with clusters of galaxies supply a lower limit to the density parameter. Barring seemingly improbable complications, these constraints confine all allowed models to a small range of values of the density parameter and the Hubble time (*colored area*); all models in that range describe a universe that is open, infinite and perpetually expanding.

# 11

# The Search for Extraterrestrial Intelligence

by Carl Sagan and Frank Drake
*May 1975*

*There can be little doubt that civilizations more advanced than the earth's exist elsewhere in the universe. The probabilities involved in locating one of them call for a substantial effort*

Is mankind alone in the universe? Or are there somewhere other intelligent beings looking up into their night sky from very different worlds and asking the same kind of question? Are there civilizations more advanced than ours, civilizations that have achieved interstellar communication and have established a network of linked societies throughout our galaxy? Such questions, bearing on the deepest problems of the nature and destiny of mankind, were long the exclusive province of theology and speculative fiction. Today for the first time in human history they have entered into the realm of experimental science.

From the movements of a number of nearby stars we have now detected unseen companion bodies in orbit around them that are about as massive as large planets. From our knowledge of the processes by which life arose here on the earth we know that similar processes must be fairly common throughout the universe. Since intelligence and technology have a high survival value it seems likely that primitive life forms on the planets of other stars, evolving over many billions of years, would occasionally develop intelligence, civilization and a high technology. Moreover, we on the earth now possess all the technology necessary for communicating with other civilizations in the depths of space. Indeed, we may now be standing on a threshold about to take the momentous step a planetary society takes but once: first contact with another civilization.

In our present ignorance of how common extraterrestrial life may actually be, any attempt to estimate the number of technical civilizations in our galaxy is necessarily unreliable. We do, however, have some relevant facts. There is reason to believe that solar systems are formed fairly easily and that they are abundant in the vicinity of the sun. In our own solar system, for example, there are three miniature "solar systems": the satellite systems of the planets Jupiter (with 13 moons), Saturn (with 10) and Uranus (with five). It is plain that however such systems are made, four of them formed in our immediate neighborhood.

The only technique we have at present for detecting the planetary systems of nearby stars is the study of the gravitational perturbations such planets induce in the motion of their parent star. Imagine a nearby star that over a period of decades moves measurably with respect to the background of more distant stars. Suppose it has a nonluminous companion that circles it in an orbit whose plane does not coincide with our line of sight to the star. Both the star and the companion revolve around a common center of mass. The center of mass will trace a straight line against the stellar background and thus the luminous star will trace a sinusoidal path. From the existence of the oscillation we can deduce the existence of the companion. Furthermore, from the period and amplitude of the oscillation we can calculate the period and mass of the companion. The technique is only sensitive enough, however, to detect the perturbations of a massive planet around the nearest stars.

The single star closest to the sun is Barnard's star, a rather dim red dwarf about six light-years away. (Although Alpha Centauri is closer, it is a member of a triple-star system.) Observations made by Peter van de Kamp of the Sproul Observatory at Swarthmore College over a period of 40 years suggest that Barnard's star is accompanied by at least two dark companions, each with about the mass of Jupiter.

There is still some controversy over his conclusion, however, because the observations are very difficult to make. Perhaps even more interesting is the fact that of the dozen or so single stars nearest the sun nearly half appear to have dark companions with a mass between one and 10 times the mass of Jupiter. In

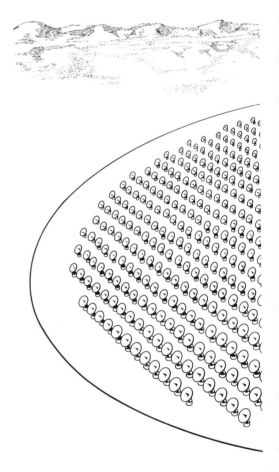

"CYCLOPS," an array of 1,500 radio antennas each 100 meters in diameter, is one system that has been proposed as a tool for detecting signals from extraterrestrial civ-

addition many theoretical studies of the formation of planetary systems out of contracting clouds of interstellar gas and dust imply that the birth of planets frequently if not inevitably accompanies the birth of stars.

We know that the master molecules of living organisms on the earth are the proteins and the nucleic acids. The proteins are built up of amino acids and the nucleic acids are built up of nucleotides. The earth's primordial atmosphere was, like the rest of the universe, rich in hydrogen and in hydrogen compounds. When molecular hydrogen ($H_2$), methane ($CH_4$), ammonia ($NH_3$) and water ($H_2O$) are mixed together in the presence of virtually any intermittent source of energy capable of breaking chemical bonds, the result is a remarkably high yield of amino acids and the sugars and nitrogenous bases that are the chemical constituents of the nucleotides. For example, from laboratory experiments we can determine the amount of amino acids produced per photon of ultraviolet radiation, and from our knowledge of

stellar evolution we can calculate the amount of ultraviolet radiation emitted by the sun over the first billion years of the existence of the earth. Those two rates enable us to compute the total amount of amino acids that were formed on the primitive earth. Amino acids also break down spontaneously at a rate that is dependent on the ambient temperature. Hence we can calculate their steady-state abundance at the time of the origin of life. If amino acids in that abundance were mixed into the oceans of today, the result would be a 1 percent solution of amino acids. That is approximately the concentration of amino acids in the better brands of canned chicken bouillon, a solution that is alleged to be capable of sustaining life.

The origin of life is not the same as the origin of its constituent building blocks, but laboratory studies on the linking of amino acids into molecules resembling proteins and on the linking of nucleotides into molecules resembling nucleic acids are progressing well. In-

vestigations of how short chains of nucleic acids replicate themselves in vitro have even provided clues to primitive genetic codes for translating nucleic acid information into protein information, systems that could have preceded the elaborate machinery of ribosomes and activating enzymes with which cells now manufacture protein.

The laboratory experiments also yield a large amount of a brownish polymer that seems to consist mainly of long hydrocarbon chains. The spectroscopic properties of the polymer are similar to those of the reddish clouds on Jupiter, Saturn and Titan, the largest satellite of Saturn. Since the atmospheres of these objects are rich in hydrogen and are similar to the atmosphere of the primitive earth, the coincidence is not surprising. It is nonetheless remarkable. Jupiter, Saturn and Titan may be vast planetary laboratories engaged in prebiological organic chemistry.

Other evidence on the origin of life comes from the geological record of the earth. Thin sections of sedimentary rocks

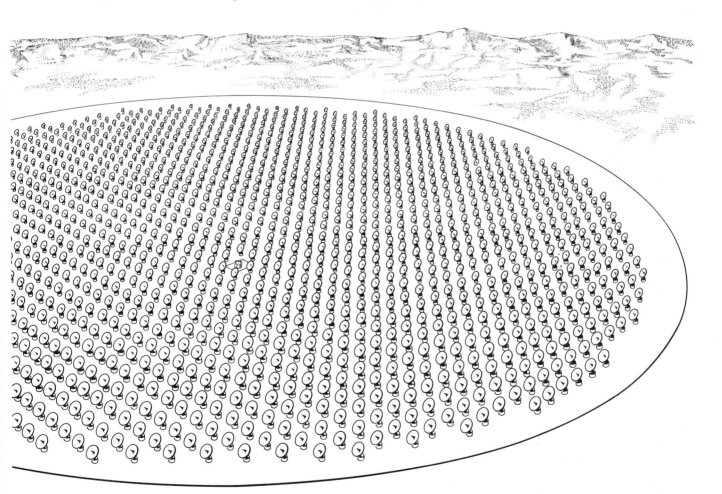

ilizations. The individual antennas would be connected to one another and to a large computer system. The effective signal-collecting area of the system would be hundreds of times greater than that of any existing radio telescope and would be capable of detecting even such relatively weak signals as the internal radio-frequency communications of a civilization as far away as several hundred light-years. Control building shown in center of array includes observatory with telescope operating at visible wavelengths.

between 2.7 and 3.5 billion years old reveal the presence of small inclusions a hundredth of a millimeter in diameter. These inclusions have been identified by Elso S. Barghoorn of Harvard University and J. William Schopf of the University of California at Los Angeles as bacteria and blue-green algae. Bacteria and blue-green algae are evolved organisms and must themselves be the beneficiaries of a long evolutionary history. There are no rocks on the earth or on the moon, however, that are more than four billion years old; before that time the surface of both bodies is believed to have melted in the final stages of their accretion. Thus the time available for the origin of life seems to have been short: a few hundred million years at the most. Since life originated on the earth in a span much shorter than the present age of the earth, we have additional evidence that the origin of life has a high probability, at least on planets with an abundant supply of hydrogen-rich gases, liquid water and

sources of energy. Since those conditions are common throughout the universe, life may also be common.

Until we have discovered at least one example of extraterrestrial life, however, that conclusion cannot be considered secure. Such an investigation is one of the objectives of the Viking mission, which is scheduled to land a vehicle on the surface of Mars in the summer of 1976, a vehicle that will conduct the first rigorous search for life on another planet. The *Viking* lander carries three separate experiments on the metabolism of hypothetical Martian microorganisms, one experiment on the organic chemistry of the Martian surface material and a camera system that might just conceivably detect macroscopic organisms if they exist.

Intelligence and technology have developed on the earth about halfway through the stable period in the lifetime of the sun. There are obvious selective

advantages to intelligence and technology, at least up to the present evolutionary stage when technology also brings the threats of ecological catastrophes, the exhaustion of natural resources and nuclear war. Barring such disasters, the physical environment of the earth will remain stable for many more billions of years. It is possible that the number of individual steps required for the evolution of intelligence and technology is so large and improbable that not all inhabited planets evolve technical civilizations. It is also possible—some would say likely—that civilizations tend to destroy themselves at about our level of technological development. On the other hand, if there are 100 billion suitable planets in our galaxy, if the origin of life is highly probable, if there are billions of years of evolution available on each such planet and if even a small fraction of technical civilizations pass safely through the early stages of technological adolescence, the number of technological civi-

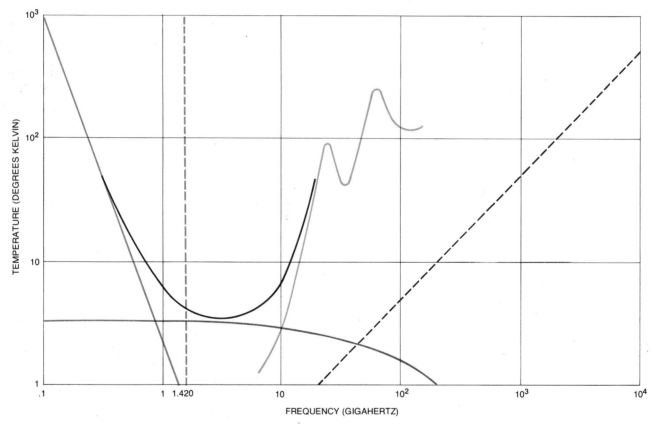

**RADIO SPECTRUM** of the sky as it is seen from the earth is quite noisy. Any radio telescope picks up the three-degree-Kelvin background radiation (*gray line*), the remnant of the primordial fireball of the "big bang." The background radiation begins to fall off at about 60 gigahertz (billion cycles per second). At that frequency the quantum noise associated with all electromagnetic radiation (*broken black line*) begins to predominate, and the total noise level rises. Noise from within our galaxy (*dark colored line*) is due mainly to synchrotron radiation emitted by particles spiraling in around the lines of force in magnetic fields. Together these three sources of noise define a broad quiet region in the radio spectrum, between about one gigahertz and 100 gigahertz, that would be nearly the same for observers in the neighborhood of the sun and observers in similar regions of the galaxy. The earth's atmosphere is also a source of noise (*light colored line*) because molecules of water and oxygen absorb and reradiate energy at 22 gigahertz and 60 gigahertz. All sources of noise added together yield the curve in black, representing the total sky noise detected on the earth. The broken vertical line in color is frequency of spin-flip of the electron in un-ionized hydrogen atom at frequency of 1.420 gigahertz.

lizations in the galaxy today might be very large.

It is obviously a highly uncertain exercise to attempt to estimate the number of such civilizations. The opinions of those who have considered the problem differ significantly. Our best guess is that there are a million civilizations in our galaxy at or beyond the earth's present level of technological development. If they are distributed randomly through space, the distance between us and the nearest civilization should be about 300 light-years. Hence any information conveyed between the nearest civilization and our own will take a minimum of 300 years for a one-way trip and 600 years for a question and a response.

Electromagnetic radiation is the fastest and also by far the cheapest method of establishing such contact. In terms of the foreseeable technological developments on the earth, the cost per photon and the amount of absorption of radiation by interstellar gas and dust, radio waves seem to be the most efficient and economical method of interstellar communication. Interstellar space vehicles cannot be excluded a priori, but in all cases they would be a slower, more expensive and more difficult means of communication.

Since we have achieved the capability for interstellar radio communication only in the past few decades, there is virtually no chance that any civilization we come in contact with will be as backward as we are. There also seems to be no possibility of dialogue except between very long-lived and patient civilizations. In view of these circumstances, which should be common to and deducible by all the civilizations in our galaxy, it seems to us quite possible that one-way radio messages are being beamed at the earth at this moment by radio transmitters on planets in orbit around other stars.

To intercept such signals we must guess or deduce the frequency at which the signal is being sent, the width of the frequency band, the type of modulation and the star transmitting the message. Although the correct guesses are not easy to make, they are not as hard as they might seem.

Most of the astronomical radio spectrum is quite noisy [see *illustration on opposite page*]. There are contributions from interstellar matter, from the three-degree-Kelvin background radiation left over from the early history of the universe, from noise that is fundamentally associated with the operation of any detector and from the absorption of radia-

| INVESTIGATOR | OBSERVATORY | DATE | FREQUENCY OR WAVELENGTH | TARGETS |
|---|---|---|---|---|
| DRAKE | N.R.A.O. | 1960 | 1,420 MEGAHERTZ | EPSILON ERIDANI TAU CETI |
| TROITSKY | GORKY | 1968 | 21 AND 30 CENTIMETERS | 12 NEARBY SUNLIKE STARS |
| VERSCHUUR | N.R.A.O. | 1972 | 1,420 MEGAHERTZ | 10 NEARBY STARS |
| TROITSKY | EURASIAN NETWORK, GORKY | 1972 TO PRESENT | 16, 30 AND 50 CENTIMETERS | PULSED SIGNALS FROM ENTIRE SKY |
| ZUCKERMAN PALMER | N.R.A.O. | 1972 TO PRESENT | 1,420 MEGAHERTZ | ~600 NEARBY SUNLIKE STARS |
| KARDASHEV | EURASIAN NETWORK, I.C.R. | 1972 TO PRESENT | SEVERAL | PULSED SIGNALS FROM ENTIRE SKY |
| BRIDLE FELDMAN | A.R.O. | 1974 TO PRESENT | 22.2 GIGAHERTZ | SEVERAL NEARBY STARS |
| DRAKE SAGAN | ARECIBO | 1975 (IN PROGRESS) | 1,420, 1,653 AND 2,380 MEGAHERTZ | SEVERAL NEARBY GALAXIES |

**ATTEMPTS TO DETECT SIGNALS** beamed toward the earth by other civilizations have so far been unsuccessful, but the number of stars that have been examined is less than .1 percent of the number that would have to be investigated if there were to be a reasonable statistical chance of discovering one extraterrestrial civilization. "N.R.A.O." is the National Radio Astronomy Observatory in Green Bank, W.Va.; "Gorky" is the 45-foot antenna at Gorky University in the U.S.S.R.; "Eurasian network" is a network of omnidirectional antennas in the U.S.S.R. that is being operated jointly by V. S. Troitsky of Gorky University and N. S. Kardashev of the Institute for Cosmic Research (I.C.R.) of the Academy of Sciences of the U.S.S.R.; "A.R.O." is the Algonquin Radio Observatory at Algonquin Park in Canada; "Arecibo" is 1,000-foot radio-radar antenna at Arecibo Observatory in Puerto Rico.

tion by the earth's atmosphere. This last source of noise can be avoided by placing a radio telescope in space. The other sources we must live with and so must any other civilization.

There is, however, a pronounced minimum in the radio-noise spectrum. Lying at the minimum or near it are several natural frequencies that should be discernible by all scientifically advanced societies. They are the resonant frequencies emitted by the more abundant molecules and free radicals in interstellar space. Perhaps the most obvious of these resonances is the frequency of 1,420 megahertz (millions of cycles per second). That frequency is emitted when the spinning electron in an atom of hydrogen spontaneously flips over so that its direction of spin is opposite to that of the proton comprising the nucleus of the hydrogen atom. The frequency of the spin-flip transition of hydrogen at 1,420 megahertz was first suggested as a channel for interstellar communication

in 1959 by Philip Morrison and Giuseppe Cocconi. Such a channel may be too noisy for communication precisely because hydrogen, the most abundant interstellar gas, absorbs and emits radiation at that frequency. The number of other plausible and available communication channels is not large, so that determining the right one should not be too difficult.

We cannot use a similar logic to guess the bandwidth that might be used in interstellar communication. The narrower the bandwidth is, the farther a signal can be transmitted before it becomes too weak for detection. On the other hand, the narrower the bandwidth is, the less information the signal can carry. A compromise is therefore required between the desire to send a signal the maximum distance and the desire to communicate the maximum amount of information. Perhaps simple signals with narrow bandwidths are sent to enhance the probability of the signals' being received. Perhaps information-rich signals with broad bandwidths are sent in order to achieve rapid and extensive communication. The broad-bandwidth signals would be intended for those enlightened civilizations that have invested major resources in large receiving systems.

When we actually search for signals, it is not necessary to guess the exact bandwidth, only to guess the minimum bandwidth. It is possible to communicate on many adjacent narrow bands at once. Each such channel can be studied individually, and the data from several adjacent channels can be combined to yield the equivalent of a wider channel without any loss of information or sensitivity. The procedure is relatively easy with the aid of a computer; it is in fact routinely employed in studies of pulsars. In any event we should observe the maximum number of channels because of the possibility that the transmitting civilization is not broadcasting on one of the "natural" frequencies such as 1,420 megahertz.

We do not, of course, know now which star we should listen to. The most conservative approach is to turn our receivers to stars that are rather similar to the sun, beginning with the nearest. Two nearby stars, Epsilon Eridani and Tau Ceti, both about 12 light-years away, were the candidates for Project Ozma, the first search with a radio telescope for extraterrestrial intelligence, conducted by one of us (Drake) in 1960. Project Ozma, named after the ruler of Oz in L. Frank Baum's children's stories, was "on the air" for four weeks at 1,420 megahertz. The results were negative. Since then there have been a number of other studies. In spite of some false alarms to the contrary, none has been successful. The lack of success is not unexpected. If there are a million technical civilizations in a galaxy of some 200 billion stars, we must turn our receivers to 200,000 stars before we have a fair statistical chance of detecting a single extraterrestrial message. So far we have listened to only a few more than 200 stars. In other words, we have mounted only .1 percent of the required effort.

Our present technology is entirely adequate for both transmitting and receiving messages across immense interstellar distances. For example, if the 1,000-foot radio telescope at the Arecibo Observatory in Puerto Rico were to transmit information at the rate of one bit (binary digit) per second with a bandwidth of one hertz, the signal could

be received by an identical radio telescope anywhere in the galaxy. By the same token, the Arecibo telescope could detect a similar signal transmitted from a distance hundreds of times greater than our estimate of 300 light-years to the nearest extraterrestrial civilization.

A search of hundreds of thousands of stars in the hope of detecting one message would require remarkable dedication and would probably take several decades. It seems unlikely that any existing major radio telescope would be given over to such an intensive program to the exclusion of its usual work. The construction of one radio telescope or more that would be devoted perhaps half-time to the search seems to be the

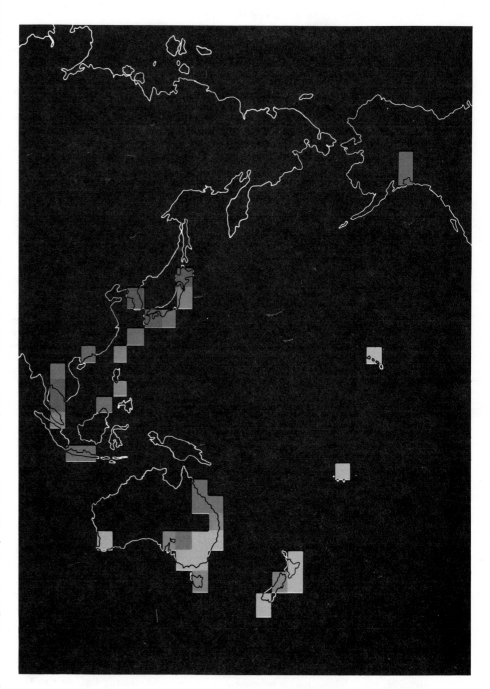

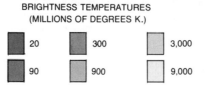

BRIGHTNESS TEMPERATURES
(MILLIONS OF DEGREES K.)

| | | |
|---|---|---|
| 20 | 300 | 3,000 |
| 90 | 900 | 9,000 |

EARTH IS BRIGHT at the frequencies between 40 and 220 megahertz because of the radiation from FM radio and VHF television broadcasts. The power radiated by the stations is shown averaged over squares five degrees in longitude by five degrees in latitude. The radio brightness is equivalent to

only practical method of seeking out extraterrestrial intelligence in a serious way. The cost would be some tens of millions of dollars.

So far we have been discussing the reception of messages that a civilization would intentionally transmit to the earth. An alternative possibility is that we might try to "eavesdrop" on the radio traffic an extraterrestrial civilization employs for its own purposes. Such radio traffic could be readily apparent. On the earth, for example, a new radar system employed with the telescope at the Arecibo Observatory for planetary studies emits a narrow-bandwidth signal that, if it were detected from another star, would be between a million and 10 billion times brighter than the sun at the same frequency. In addition, because of radio and television transmission, the earth is extremely bright at wavelengths of about a meter [see illustration on these two pages]. If the planets of other civilizations have a radio brightness comparable to the earth's

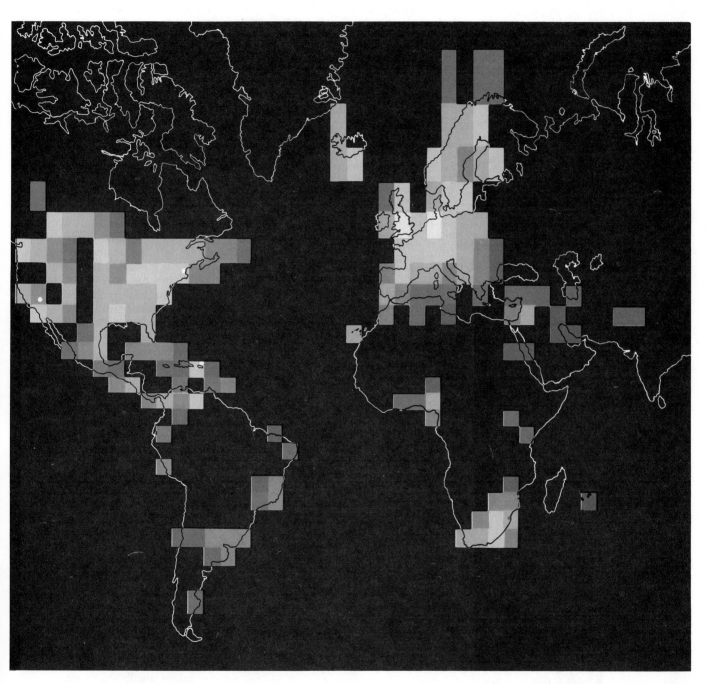

the temperature to which each area on the earth would have to be raised in order to produce the actual radio emission observed. The three brightest areas are the locations of three particularly powerful radar systems: the radio-radar antenna of the Haystack Observatory in Massachusetts, operating at a wavelength of 3.75 centimeters and giving a brightness temperature of $2.3 \times 10^{20}$ degrees K., the 1,000-foot radio-radar antenna of the Arecibo Observatory, operating at a wavelength of 12.6 centimeters and giving a brightness temperature of $1.4 \times 10^{21}$ degrees K., and the 210-foot antenna of the Jet Propulsion Laboratory at Goldstone, Calif., operating at a wavelength of 12.6 centimeters and giving a brightness temperature of $6.2 \times 10^{19}$ degrees. Systems radiate so much power that at those wavelengths and in the direction of their beam they are brighter than the sun and should be detectable over interstellar distances.

from television transmission alone, they should be detectable. Because of the complexity of the signals and the fact that they are not beamed specifically at the earth, however, the receiver we would need in order to eavesdrop would have to be much more elaborate and sensitive than any radio-telescope system we now possess.

One such system has been devised in a preliminary way by Bernard M. Oliver of the Hewlett-Packard Company, who directed a study sponsored by the Ames Research Center of the National Aeronautics and Space Administration. The system, known as Cyclops, would consist of an enormous radio telescope connected to a complex computer system. The computer system would be designed particularly to search through the data

from the telescope for signals bearing the mark of intelligence, to combine numerous adjacent channels in order to construct signals of various effective bandwidths and to present the results of the automatic analyses for all conceivable forms of interstellar radio communication in a way that would be intelligible to the project scientists.

To construct a radio telescope of enormous aperture as a single antenna would be prohibitively expensive. The Cyclops system would instead capitalize on our ability to connect many individual antennas to act in unison. This concept is already the basis of the Very Large Array now under construction in New Mexico. The Very Large Array consists of 27 antennas, each 82 feet in

diameter, arranged in a Y-shaped pattern whose three arms are each 10 miles long. The Cyclops system would be much larger. Its current design calls for 1,500 antennas each 100 meters in diameter, all electronically connected to one another and to the computer system. The array would be as compact as possible but would cover perhaps 25 square miles.

The effective signal-collecting area of the system would be hundreds of times the area of any existing radio telescope, and it would be capable of detecting even relatively weak signals such as television transmissions from civilizations several hundred light-years away. Moreover, it would be the instrument par excellence for receiving signals specifically directed at the earth. One of the greatest virtues of the Cyclops system is that no technological advances would be required in order to build it. The necessary electronic and computer techniques are already well developed. We would need only to build a vast number of items we already build well. The Cyclops system not only would have enormous power for searching for extraterrestrial intelligence but also would be an extraordinary tool for radar studies of the bodies in the solar system, for traditional radio astronomy outside the solar system and for the tracking of space vehicles to distances beyond the reach of present receivers.

The estimated cost of the Cyclops system, ranging up to $10 billion, may make it prohibitively expensive for the time being. Moreover, the argument in favor of eavesdropping is not completely persuasive. Half a century ago, before radio transmissions were commonplace, the earth was quiet at radio wavelengths. Half a century from now, because of the development of cable television and communication satellites that relay signals in a narrow beam, the earth may again be quiet. Thus perhaps for only a century out of billions of years do planets such as the earth appear remarkably bright at radio wavelengths. The odds of our discovering a civilization during that short period in its history may not be good enough to justify the construction of a system such as Cyclops. It may well be that throughout the universe beings usually detect evidence of extraterrestrial intelligence with more traditional radio telescopes. It nonetheless seems clear that our own chances of finding extraterrestrial intelligence will improve if we consciously attempt to find it.

How could we be sure that a particu-

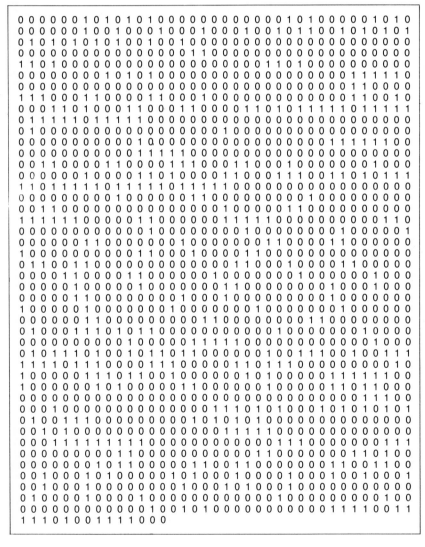

ARECIBO MESSAGE IN BINARY CODE was transmitted in 1974 toward the Great Cluster in Hercules from the 1,000-foot antenna at Arecibo. The message is decoded by breaking up the characters into 73 consecutive groups of 23 characters each and arranging the groups in sequence one under the other, reading right to left and then top to bottom. The result is a visual message (see *illustration on opposite page*) that can be more easily interpreted by making each 0 of binary code represent a white square and each 1 a black square.

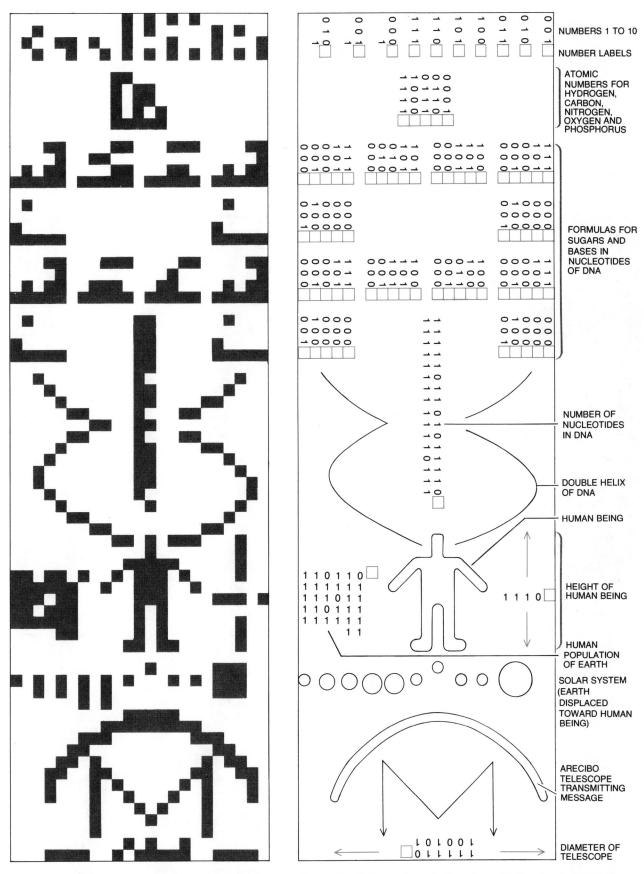

**ARECIBO MESSAGE IN PICTURES** and accompanying translation shows the binary version of the message decoded. Each number that is used is marked with a label that indicates its start. When all the digits of a number cannot be fitted into one line, the digits for which there is no room are written under the least significant digit. (The message must be oriented in three different ways for all the numbers shown to be read.) The chemical formulas are those for the components of the DNA molecule: the phosphate group, the deoxyribose sugar and the organic bases thymine, adenine, guanine and cytosine. Both the height of the human being and the diameter of the telescope are given in units of the wavelength that is used to transmit the message: 12.6 centimeters.

lar radio signal was deliberately sent by an intelligent being? It is easy to design a message that is unambiguously artificial. The first 30 prime numbers, for example, would be difficult to ascribe to some natural astrophysical phenomenon. A simple message of this kind might be a beacon or announcement signal. A subsequent informative message could have many forms and could consist of an enormous number of bits. One method of transmitting information, beginning simply and progressing to more elaborate concepts, is pictures [see illustration on preceding page].

One final approach in the search for extraterrestrial intelligence deserves mention. If there are indeed civilizations thousands or millions of years more advanced than ours, it is entirely possible that they could beam radio communications over immense distances, perhaps even over the distances of intergalactic space. We do not know how many advanced civilizations there might be compared with the number of more primitive earthlike civilizations, but many of these older civilizations are bound to be in galaxies older than our own. For this reason the most readily detectable radio signals from another civilization may come from outside our galaxy. The relatively small number of such extragalactic transmitters might be more than compensated for by the greater strength

THOUSAND-FOOT ANTENNA of the radio-radar system at the Arecibo Observatory is made of perforated aluminum panels whose spherical shape is accurate to within 1/8 inch over the antenna's entire area of 20 acres. The triangular structure suspended above the antenna holds the receiver and the transmitter for the system. Control rooms and office buildings are to lower right of antenna.

of their signals. At the appropriate frequency they could even be the brightest radio signals in the sky. Therefore an alternative to examining the nearest stars of the same spectral type as the sun is to examine the nearest galaxies. Spiral galaxies such as the Great Nebula in Andromeda are obvious candidates, but the elliptical galaxies are much older and more highly evolved and could conceivably harbor a large number of extremely advanced civilizations.

There might be a kind of biological law decreeing that there are many paths to intelligence and high technology, and that every inhabited planet, if it is given enough time and it does not destroy itself, will arrive at a similar result. The biology on other planets is of course expected to be different from our own because of the statistical nature of the evolutionary process and the adaptability of life. The science and engineering, however, may be quite similar to ours, because any civilization engaged in interstellar radio communication, no matter where it exists, must contend with the same laws of physics, astronomy and radio technology that we do.

Should we be sending messages ourselves? It is obvious that we do not yet know where we might best direct them. One message has already been transmitted to the Great Cluster in Hercules by the Arecibo radio telescope, but only as a kind of symbol of the capabilities of our existing radio technology. Any radio signal we send would be detectable over interstellar distances if it is more than about 1 percent as bright as the sun at the same frequency. Actually something close to 1,000 such signals from our everyday internal communications have left the earth every second for the past two decades. This electromagnetic frontier of mankind is now some 20 light-years away, and it is moving outward at the speed of light. Its spherical wave front, expanding like a ripple from a disturbance in a pool of water and inadvertently carrying the news that human beings have achieved the capacity for interstellar discourse, envelops about 20 new stars each year.

We have also sent another kind of message: two engraved plaques that ride aboard *Pioneer 10* and *Pioneer 11*. These spacecraft, the first artifacts of mankind that will escape from the solar system, will voyage forever through our galaxy at a speed of some 10 miles per second. *Pioneer 10* was accelerated to the velocity of escape from the solar system by the gravitational field of Jupiter

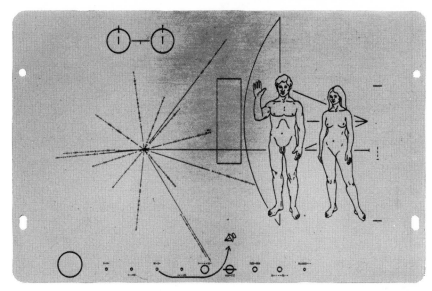

ENGRAVED PLAQUE on the *Pioneer* spacecraft to Jupiter is another message that has been dispatched beyond the solar system. Meaning of symbols is given in text of article.

on December 3, 1973. *Pioneer 11* swung past Jupiter on December 4, 1974, and will travel on to Saturn before it is accelerated on a course to the far side of the galaxy.

Identical plaques for each vehicle were designed by us and Linda Salzman Sagan. Each plaque measures six by nine inches and is made of gold-anodized aluminum. These engraved cosmic greeting cards bear the location of the earth and the time the spacecraft was built and launched. The sun is located with respect to 14 pulsars. The precise periods of the pulsars are specified in binary code to allow them to be identified. Since pulsars are cosmic clocks that are running down at a largely constant rate, the difference in the pulsar periods at the time one of the spacecraft is recovered and the periods indicated on the plaque will enable any technically sophisticated civilization to deduce the year the vehicle was sent on its epic journey. Units of time and distance are specified in terms of the frequency of the hydrogen spin-flip at 1,420 megahertz. In order to identify the exact location of the spacecraft's launch a diagram of the solar system is given. The trajectory of the spacecraft is shown as it leaves the third planet, the earth, and swings by the fifth planet, Jupiter. (The diversion of *Pioneer 11* past Saturn had not been planned when the plaques were prepared.) Last, the plaques show images of a man and a woman of the earth in 1973. An attempt was made to give the images panracial characteristics. Their heights are shown

with respect to the spacecraft and are also given by a binary number stated in terms of the wavelength of the spectral line at 1,420 megahertz (21 centimeters).

These plaques are destined to be the longest-lived works of mankind. They will survive virtually unchanged for hundreds of millions, perhaps billions, of years in space. When plate tectonics has completely rearranged the continents, when all the present landforms on the earth have been ground down, when civilization has been profoundly transformed and when human beings may have evolved into some other kind of organism, these plaques will still exist. They will show that in the year we called 1973 there were organisms, portrayed on the plaques, that cared enough about their place in the hierarchy of all intelligent beings to share knowledge about themselves with others.

How much do we care? Enough to devote an appreciable effort with existing telescopes to search for life elsewhere in the universe? Enough to take a major step such as Project Cyclops that offers a greater chance of carrying us across the threshold, to finally communicate with a variety of extraterrestrial beings who, if they exist, would inevitably enrich mankind beyond imagination? The real question is not how, because we know how; the question is when. If enough of the beings of the earth cared, the threshold might be crossed within the lifetime of most of those alive today.

# BIOGRAPHICAL NOTES
# AND BIBLIOGRAPHIES

## 1. The Red-Shift

### *The Author*

ALLAN R. SANDAGE is astronomer at the Hale (formerly Mount Wilson and Palomar) Observatories. He attended Miami University in Ohio, entered the Navy in 1945, and completed his undergraduate work at the University of Illinois in 1948. He acquired his Ph.D. in astronomy at the California Institute of Technology in 1953, having meanwhile joined the staff of Mount Wilson and Palomar. In 1960 he was awarded the Helen Warner prize of the American Astronomical Society and in 1963 he won the Eddington Medal of the Royal Astronomical Society. In 1960 Sandage and Thomas A. Matthews were the first to isolate the quasars.

### *Bibliography*

COSMOLOGY: A SEARCH FOR TWO NUMBERS. Allan R. Sandage. *Physics Today,* vol. 23, pages 34–41; February 1970.

THE REALM OF THE NEBULAE. Edwin Hubble. Yale University Press, 1936.

RED-SHIFTS AND MAGNITUDES OF EXTRA-GALACTIC NEBULAE. M. L. Humason, N. U. Mayall, and A. R. Sandage. *Astronomical Journal,* vol. 61, pages 97–162; April 1956.

THE TIME SCALE FOR CREATION. Allan R. Sandage. In *Galaxies and the Universe,* edited by Lodewijk Woltjer. Columbia University Press, 1968.

STEPS TOWARD THE HUBBLE CONSTANT, I: CALIBRATION OF THE LINEAR SIZES OF EXTRAGALACTIC H II REGIONS. Allan Sandage and G. A. Tammann. *Astrophysical Journal,* vol. 190, pages 525–538; June 15, 1974.

STEPS TOWARD THE HUBBLE CONSTANT, IV: GALAXIES IN THE GENERAL FIELD LEADING TO A CALIBRATION OF THE GALAXY LUMINOSITY CLASSES AND A FIRST HINT OF THE VALUE OF $H_0$. Allan Sandage and G. A. Tammann. *Astrophysical Journal,* vol. 194, pages 559–568; December 15, 1974.

## 2. The Evolutionary Universe

### *The Author*

GEORGE GAMOW was professor of physics at the University of Colorado at the time of his death in 1968; he had just accepted that post in 1956, when this article was first published. Born in Odessa, Russia, in 1904, he studied nuclear physics at the University of Leningrad, where he received his doctoral degree in 1928. He also studied at the University of Copenhagen under Niels Bohr and at the University of Cambridge under Ernest Rutherford. In 1934 Gamow emigrated to the U.S. For over 20 years he was professor of physics at George Washington University. During this period Gamow found his interest turning from the atomic nucleus to astrophysics, the theory of the expanding universe and later to fundamental problems of biology, including molecular genetics and the synthesis of proteins. A quixotic streak, which enlivened his many popular books and articles, sometimes extended to his most serious scientific publications. On one occasion when he and Ralph A. Alpher were preparing a paper, they invited Hans Bethe of Cornell University to collaborate with them. The paper, which happened to concern the beginning of the universe, was therefore most appropriately authored by Alpher, Bethe and Gamow.

*Bibliography*

THE AGE OF CREATION. Allan R. Sandage. *Science Year 1968*, pages 56–69.

THE CREATION OF THE UNIVERSE. George Gamow. Viking, 1952.

FACT AND THEORY IN COSMOLOGY. G. C. McVittie. Eyre and Spottiswoode, 1961.

THE MYSTERY OF THE EXPANDING UNIVERSE. W. B. Bonner. Macmillan, 1964.

THE PRIMEVAL ATOM. Georges Lemaitre. D. Van Nostrand, 1950.

RELATIVISTIC COSMOLOGY. Wolfgang Rindler. *Physics Today*, vol. 20, pages 23–31; November 1967.

AN UNBOUND UNIVERSE? J. Richard Gott III, James E. Gunn, David N. Schramm, and Beatrice M. Tinsley. *Astrophysical Journal*, vol. 194, pages 543–553; December 15, 1974.

# 3. The Curvature of Space in a Finite Universe

*The Author*

J. J. CALLAHAN is associate professor of mathematics at Smith College. A 1962 graduate of Marist College, he went on to obtain his Ph.D. in mathematics from New York University. He was a Benjamin Peirce lecturer at Harvard University for three years before joining the faculty at Smith in 1970. His primary mathematical interests, he notes, are "differential analysis and catastrophe theory and its applications. The themes of this article—geometry and the history of mathematics—are things I am interested in teaching, but I am not a specialist in them." The article developed, he adds, "out of an attempt to explain Einstein's concept of a finite but unbounded space to my nonscientific colleagues at Smith. They found it tough going, and some simply dismissed a finite universe as impossible, because Kant had done so when he studied the question 300 years ago. A sabbatical this past year at the University of Warwick gave me a chance to read what Kant said about space. He does indeed epitomize the commonsense view, which Einstein (and Riemann before him) shattered. I was unable, however, to find any satisfactory explanation of just how the old and the new ideas fit together, so I attempted one myself."

*Bibliography*

EINSTEIN'S GENERAL THEORY OF RELATIVITY. Max Born in *Einstein's Theory of Relativity*. Dover Publications, Inc., 1962.

PARADISE. Dante Alighieri, translated by Dorothy L. Sayers and Barbara Reynolds in *The Comedy of Dante Alighieri*. Penguin Books, 1962.

MATHEMATICAL THOUGHT FROM ANCIENT TO MODERN TIMES. Morris Kline. Oxford University Press, 1972.

GRAVITATION. Charles W. Misner, Kip S. Thorne and John Archibald Wheeler. W. H. Freeman and Company, 1973.

# 4. Cosmology before and after Quasars

*The Author*

DENNIS W. SCIAMA is professor in the Department of Astrophysics at Oxford University. In 1967, when he wrote this review, he held a Peterhouse Fellowship and Lectureship in the Department of Applied Mathematics and Theoretical Physics, Cambridge University. There he built up a group of graduate and post-doctoral students working on general relativity, cosmology, and astrophysics. Born in Manchester, England, in 1926, he was a student of the great theoretical physicist P. A. M. Dirac, and received his Ph.D. degree from Cambridge University in 1952. In that same year he obtained a Research Fellowship at Trinity College, Cambridge, and since then has been a member of the Institute for Advanced Study at Princeton and an Agassiz Fellow at Harvard University.

*Bibliography*

COSMOLOGY AFTER HALF A CENTURY. William H. McCrea. *Science*, vol. 160, pages 1295–1299; June 21, 1968.

THE COUNTS OF RADIO SOURCES. M. Ryle. *Annual Review of Astronomy and Astrophysics*, vol. 6, pages 249–266; 1968.

THE EVOLUTION OF THE UNIVERSE. Hong-Yee Chiu. *Science Journal*, vol. 4, pages 33–38; August 1968.

THE PHYSICAL FOUNDATIONS OF GENERAL RELATIVITY. D. W. Sciama. Doubleday Anchor, 1969.

THE STRUCTURE OF THE UNIVERSE. E. L. Schatzman. McGraw-Hill, 1968.

COSMOLOGY TODAY. William H. McCrea. *American Scientist*, vol. 58, pages 521–527; Sept.–Oct. 1970.

# 5. The Cosmic Background Radiation

*The Author*

ADRIAN WEBSTER is at the Mullard Radio Astronomy Observatory of the University of Cambridge as a research fellow of the Royal Commission of the Exhibition of 1851 and a research fellow of Clare College. He obtained his bachelor's degree in theoretical physics at Cambridge in 1967 and his Ph.D. in radio astronomy there in 1972. For the two academic years beginning in 1971 he was a research fellow at the Miller Institute for Basic Research in Science at the University of California, Berkeley.

*Bibliography*

MODERN COSMOLOGY. Dennis W. Sciama. Cambridge University Press, 1971.

OBSERVATIONAL COSMOLOGY. M. S. Longair. *Reports on Progress in Physics*, vol. 34, pages 1125–1248; 1971.

GRAVITATION AND COSMOLOGY: PRINCIPLES AND APPLICATIONS OF THE GENERAL THEORY OF RELATIVITY. Steven Weinberg. Wiley, 1972.

EPPUR SI MUOVE. D. W. Sciama. *Comments on Astrophysics and Space Physics*, vol. 4, pages 35–39; March–April, 1972.

## 6. The Evolution of Quasars

### The Authors

MAARTEN SCHMIDT is professor of astronomy at the California Institute of Technology; FRANCIS BELLO is associate editor of *Scientific American*. Schmidt is also a member of the staff of the Hale (formerly the Mount Wilson and Palomar) Observatories and a staff member of the Owens Valley Radio Observatory. Born in the Netherlands, he took his Ph.D. at the University of Leiden in 1956. In the same year he came to the U.S. as a Carnegie Fellow; in 1959 he moved to Cal Tech.

*Bibliography*

QUASI-STELLAR OBJECTS. Geoffrey Burbidge and Margaret Burbidge. Freeman, 1967.

SPACE DISTRIBUTION AND LUMINOSITY FUNCTIONS OF QUASI-STELLAR RADIO SOURCES. Maarten Schmidt. *Astrophysical Journal*, vol. 151, pages 393–409; February, 1968.

ON THE NATURE OF FAINT BLUE OBJECTS IN HIGH GALACTIC LATITUDES, II: SUMMARY OF PHOTOMETRIC RESULTS FOR 301 OBJECTS IN SEVEN SURVEY FIELDS. Allan Sandage and Willem J. Luyten. *Astrophysical Journal*, vol. 155, pages 913–918; March, 1969.

SPACE DISTRIBUTION AND THE LUMINOSITY FUNCTIONS OF QUASARS. Maarten Schmidt, *Astrophysical Journal*, vol. 162, pages 371–379; November, 1970.

THE REDSHIFT CONTROVERSY. George B. Field, Halton Arp, and John N. Bahcall. Benjamin, 1973.

## 7. The Origin of Galaxies

### The Authors

MARTIN J. REES is Plumian Professor of Astronomy and Experimental Philosophy at the University of Cambridge. His master's degree in mathematics and Ph.D. in astrophysics are both from Cambridge. His interests include cosmology, diffuse matter in space, and theoretical radio astronomy. JOSEPH SILK is Professor of Astronomy at the University of California, Berkeley. He was a Cambridge undergraduate and obtained his Ph.D. at Harvard University.

*Bibliography*

THE BLACK-BODY RADIATION CONTENT OF THE UNIVERSE AND THE FORMATION OF GALAXIES. P. J. E. Peebles. *Astrophysical Journal*, vol. 142, No. 4, pages 1317–1326; November 15, 1965.

THE CASE FOR A HIERARCHICAL COSMOLOGY. G. de Vaucouleurs. *Science*, vol. 167, pages 1203–1213; February 27, 1970.

COSMIC BLACK-BODY RADIATION AND GALAXY FORMATION. Joseph Silk. *Astrophysical Journal*, vol. 151, pages 459–471; February 1968.

THE FORMATION AND EARLY DYNAMICAL HISTORY OF GALAXIES. G. B. Field. In *Stars and Stellar Systems, Vol. 9: Galaxies and the Universe*, edited by A. Sandage and M. Sandage. University of Chicago Press, 1976.

THE FORMATION OF STARS AND GALAXIES: UNIFIED HYPOTHESES. David Layzer. In *Annual Review of Astronomy and Astrophysics*, Vol. 2, edited by Leo Goldberg, Armin J. Deutsch, and David Layzer. Annual Review, 1964.

SOME CURRENT PROBLEMS IN GALAXY FORMATION. M. J. Rees. In *Italian Physical Society: Proceedings of the International School of Physics "Enrico Fermi," Course 47: General Relativity and Cosmology*, edited by B. K. Sachs. Academic Press, 1971.

## 8. The Search for Black Holes

### The Author

KIP S. THORNE is professor of theoretical physics at the California Institute of Technology and adjunct professor of physics at the University of Utah. He graduated from Cal Tech in 1962 and received his master's (1963) and Ph.D. (1965) degrees from Princeton University.

*Bibliography*

BEYOND THE BLACK HOLE. John A. Wheeler. *Science Year 1973*, pages 76–89.

GRAVITATION. Charles W. Misner, Kip S. Thorne, and John Archibald Wheeler. Freeman, 1973.

BLACK HOLE EXPLOSIONS? S. W. Hawking *Nature*, vol. 248, pages 30–31; March 1, 1974.

## 9. The Quantum Mechanics of Black Holes

*The Author*

S. W. HAWKING is a theoretical physicist at the University of Cambridge. He was born in Oxford in 1942 and was graduated from the University of Oxford in 1962. He did his graduate work at Cambridge on general relativity, working under the direction of D. W. Sciama. He is currently a fellow of Gonville and Caius College at Cambridge and reader in gravitational physics in the university's department of applied mathematics and theoretical physics. In 1974–1975 he was Sherman Fairchild Distinguished Scholar at the California Institute of Technology. A Fellow of the Royal Society, he has received a number of honors in the past two years, including the Eddington Medal of the Royal Astronomical Society and the Dannie Heineman Prize for Mathematical Physics of the American Physical Society and the American Institute of Physics. Since 1962, when he began his graduate work at Cambridge, Hawkins has suffered from a progressive nervous disease that has confined him to a wheelchair for the past seven years. "Fortunately," he writes, "theoretical physics is one of the few fields in which this is not a serious handicap."

*Bibliography*

THE FOUR LAWS OF BLACK HOLE MECHANICS. J. M. Bardeen, B. Carter and S. W. Hawking in *Communications in Mathematical Physics*, vol. 31, no. 2, pages 161–170; 1973.

BLACK HOLES AND ENTROPY. Jacob D. Bekenstein in *Physical Review D*, vol. 7, no. 8, pages 2333–2346; April 15, 1973.

PARTICLE CREATION BY BLACK HOLES. S. W. Hawking in *Communications in Mathematical Physics*, vol. 43, no. 3, pages 199–220; 1975.

BLACK HOLES AND THERMODYNAMICS. S. W. Hawking in *Physical Review D*, vol. 13, no. 2, pages 191–197; January 15, 1976.

## 10. Will the Universe Expand Forever?

*The Authors*

J. RICHARD GOTT III, JAMES E. GUNN, DAVID N. SCHRAMM and BEATRICE M. TINSLEY began their collaboration in 1974, while Gott and Gunn were at the California Institute of Technology and Schramm and Tinsley were at the University of Texas at Austin. The key connection came when Gott traveled to Austin to give a talk and the four realized, in Tinsley's words, "that pieces of evidence from our various research specialties could be fitted together into a surprisingly clear case for an open universe. There were several subsequent visits between Texas and California and many long telephone conversations while the work on our joint paper for *The Astrophysical Journal* was in progress, but it was not until the summer of 1975 that all of us actually got together. Then, at the Institute of Astronomy of the University of Cambridge, we used the material from our published paper to write (with the literary assistance of Rosemary W. Gunn) this article for *Scientific American*." Gott, who was graduated *summa cum laude* from Harvard University in 1969 and obtained his Ph.D. in astrophysics from Princeton University in 1972, was recently appointed to the Princeton faculty. Gunn, a graduate of Rice University, received his Ph.D. from Cal Tech in 1965. He is now professor of astronomy at Cal Tech and a staff member of the Hale Observatories. Schramm acquired his undergraduate degree from the Massachusetts Institute of Technology in 1963 and his doctorate from Cal Tech in 1971. (That same year he won the national Greco-Roman wrestling championship.) He is currently associate professor of astronomy and astrophysics at the Enrico Fermi Institute of the University of Chicago. Tinsley has an M.S. in physics from the University of Canterbury in New Zealand (1963) and a Ph.D. in astronomy from the University of Texas (1967). She is now associate professor of astronomy at Yale University.

*Bibliography*

MODERN COSMOLOGY. D. W. Sciama. Cambridge University Press, 1971.

THE AGE OF THE ELEMENTS. David N. Schramm in *Scientific American*, vol. 230, no. 1, pages 69–77; January, 1974.

WHAT CAN DEUTERIUM TELL US? David N. Schramm and Robert V. Wagoner in *Physics Today*, vol. 27, no. 12, pages 41–47; December, 1974.

AN UNBOUND UNIVERSE? J. Richard Gott III, James E. Gunn, David N. Schramm and Beatrice M. Tinsley in *The Astrophysical Journal*, vol. 194, pages 543–553; December 15, 1974.

## 11. The Search for Extraterrestrial Intelligence

*The Authors*

CARL SAGAN and FRANK DRAKE are professors of astronomy at Cornell University, where Sagan is director of the Laboratory for Planetary Studies and Drake is director of the National Astronomy and Ionosphere Center. "In returning from the International